THE FEEDBACK LOOP

....

Using Formative Assessment Data for Science Teaching and Learning

THE FEEDBACK LOOP

•••••••••••••••

Using Formative Assessment Data for Science Teaching and Learning

ERIN MARIE FURTAK
HOWARD M. GLASSER
ZORA M. WOLFE

National Science Teachers Association
Arlington, Virginia

National Science Teachers Association

Claire Reinburg, Director
Wendy Rubin, Managing Editor
Rachel Ledbetter, Associate Editor
Amanda O'Brien, Associate Editor
Donna Yudkin, Book Acquisitions Coordinator

ART AND DESIGN
Will Thomas Jr., Director
Rashad Muhammad, Graphic Designer

PRINTING AND PRODUCTION
Catherine Lorrain, Director

NATIONAL SCIENCE TEACHERS ASSOCIATION
David L. Evans, Executive Director
David Beacom, Publisher

1840 Wilson Blvd., Arlington, VA 22201
www.nsta.org/store
For customer service inquiries, please call 800-277-5300.

Copyright © 2016 by the National Science Teachers Association.
All rights reserved. Printed in the United States of America.
19 18 17 16   4 3 2 1

NSTA is committed to publishing material that promotes the best in inquiry-based science education. However, conditions of actual use may vary, and the safety procedures and practices described in this book are intended to serve only as a guide. Additional precautionary measures may be required. NSTA and the authors do not warrant or represent that the procedures and practices in this book meet any safety code or standard of federal, state, or local regulations. NSTA and the authors disclaim any liability for personal injury or damage to property arising out of or relating to the use of this book, including any of the recommendations, instructions, or materials contained therein.

PERMISSIONS
Book purchasers may photocopy, print, or e-mail up to five copies of an NSTA book chapter for personal use only; this does not include display or promotional use. Elementary, middle, and high school teachers may reproduce forms, sample documents, and single NSTA book chapters needed for classroom or noncommercial, professional-development use only. E-book buyers may download files to multiple personal devices but are prohibited from posting the files to third-party servers or websites, or from passing files to non-buyers. For additional permission to photocopy or use material electronically from this NSTA Press book, please contact the Copyright Clearance Center (CCC) (*www.copyright.com*; 978-750-8400). Please access *www.nsta.org/permissions* for further information about NSTA's rights and permissions policies.

The *Next Generation Science Standards* ("*NGSS*") were developed by twenty-six states, in collaboration with the National Research Council, the National Science Teachers Association, and the American Association for the Advancement of Science in a process managed by Achieve Inc.

Library of Congress Cataloging-in-Publication Data
Names: Furtak, Erin Marie, author. | Glasser, Howard M., 1978- author. | Wolfe, Zora M., 1975- author.
Title: The feedback loop : using formative assessment data for science teaching and learning / Erin Marie Furtak, Howard M. Glasser, Zora M. Wolfe.
Description: Arlington, VA : National Science Teachers Association, 2016. | Includes bibliographical references and index.
Identifiers: LCCN 2015042292 (print) | LCCN 2016003071 (ebook) | ISBN 9781941316146 (print) | ISBN 9781681400051 (e-book)
Subjects: LCSH: Science--Study and teaching (Elementary)--Evaluation. | Science--Study and teaching (Secondary)--Evaluation. | Science--Ability testing. | Educational tests and measurements. | Effective teaching.
Classification: LCC LB1585 .F87 2016 (print) | LCC LB1585 (ebook) | DDC 372.35/044--dc23
LC record available at *http://lccn.loc.gov/2015042292*

CONTENTS

Acknowledgments vii

About the Authors ix

Contributors xi

Introduction xiii

PART 1: Elements of the Feedback Loop

Chapter 1: Overview of the Feedback Loop 3

Chapter 2: Setting Goals 15

    A Staircase Progression in a Non-*NGSS* State

    Quick-Survey Formative Assessment With Multiple-Choice Questions and Votes
        Kelly Lubkeman

    Resource Activity 2.1: Unpacking an *NGSS* (or Other Type of) Standard 34

    Resource Activity 2.2: Defining a Staircase Learning Progression 35

Chapter 3: Designing, Selecting, and Adapting Tools 37

    Supporting Goal–Tool Alignment With the Feedback Loop

    Setting Norms for Formative Assessment Conversations

    How Do We Know If Climate Change Is Happening?

    Resource Activity 3.1: Brainstorming Features for Multicomponent
    Formative Assessment Tools 65

    Resource Activity 3.2: Evaluating the Quality of Formative Assessment Tools ... 66

    Resource Activity 3.3: Anticipation Guide 67

Chapter 4: Collecting Data 69

    Initial and Revised Models in the Feedback Loop
        Kate Henson

    Is It Heat? Is It Temperature?
        Stephan Graham

    Resource Activity 4.1: Data Collection Plan 85

Chapter 5: Making Inferences 87

    One Day, Multiple Feedback Loops
        Deb Morrison

    Resource Activity 5.1: Guide to Tracking Inferences 101

Chapter 6: Closing the Feedback Loop 103

    Feedback in an Assessment Conversation

    One Tool, Two Sources of Data

    Resource Activity 6.1: Multiple Feedback Loops 119

PART 2: Using the Feedback Loop

Chapter 7: Using the Feedback Loop to Plan and Inform Instruction 123

    Refining Practice With the Feedback Loop
      Erin Zekis

    Resource Activity 7.1: Reflecting on Previously Collected Data 137

    Resource Activity 7.2: Planning for Instruction With the Feedback Loop 139

Chapter 8: Collaborating With Colleagues 141

    Collaborative Tool Design in a Teacher Learning Community
      Erin Marie Furtak and Sara C. Heredia

    Exploring the *NGSS* Through Task Analysis
      Angela Haydel DeBarger, Christopher Harris, William Penuel, Katie Van Horne

    Resource Activity 8.1: Collaborative Feedback Loop Protocol 158

Chapter 9: Resources 163

Glossary of Key Terms 167

Index 169

ACKNOWLEDGMENTS

This book would not have been possible without the extensive support of the organizations for which we work or have worked. We are grateful for the original collaborative effort that brought us together and propelled us onto a trajectory which, nearly five years later, has resulted in this book.

Vanessa de Leon helped us draft early graphical representations and assisted with final formatting of the figures. Sara C. Heredia, Deb Morrison, Ian Renga, Jason Buell, Becky Swanson, and Kate Henson were integral partners in performing the background research that informed many of the cases and concepts in this book. Tom Furtak provided guidance in development of the solenoid example in Chapter 3, and Enrique Suarez helped develop the "ting tang tong" example in Chapter 5. Sarah Roberts read and gave thoughtful feedback on a complete version of the manuscript. As with our earlier work, this book draws on the rich data collected from Rich Shavelson and Maria Araceli Ruiz-Primo's "Romance" (among curriculum, teaching, and assessment) study from the Stanford Education Assessment Laboratory, and we appreciate the opportunity to look at those data through a fresh lens.

We are thankful for the time and effort that our practicing teacher contributors—Erin Zekis, Kate Henson, Stephan Graham, and Kelly Lubkeman—dedicated to developing vignettes to provide color and teacher voice to this manuscript. At the same time, many more teachers who are identified by pseudonyms due to human subject protections contributed cases and valuable feedback that enriched the stories. In addition, we thank the students who appear in this book, who provided vital opportunities for us to learn more about the design, enactment, and improvement of formative assessment in partnership with teachers.

Finally, we are grateful for the support of our spouses and families, who were accepting of the evenings and weekends we dedicated to this project. Erin would like to thank her husband, Dave Suss, for allowing her stay at a hotel to get the final pieces of the manuscript finished, and to thank Maia and Aidan for being understanding (someday in the future) of what Erin was doing on her computer all those days. Erin's parents, Tom and Kay Furtak, were there—as always—to listen, advise, and take care of the kids. Howard would like to thank his wife, Maggie Powers, for her amazing support and patience as Howard talked through different ideas and participated in many video chats and late-night writing sessions to draft and collaborate on this book. He also thanks his parents, brother, and other family members who encouraged him and provided advice as he navigated the process of coauthoring his first book. Zora thanks her husband, Wojtek Wolfe, for his encouragement and support throughout this project, which often included wrangling the three monkeys, Alexander, Benjamin, and Christopher, as Mommy "talked to her teacher friends."

Several funding agencies also provided financial support for the research that informed this work: the National Science Foundation; the University of Colorado Innovative Seed Grant Program; the Alexander von Humboldt Foundation; and a fourth organization, which

ACKNOWLEDGMENTS

prefers to remain anonymous, that took a risk and provided research support to Erin early in her academic career. This material is based in part on work supported by the National Science Foundation under award numbers 0095520, 0953375, and 1505527. Any opinions, findings, and conclusions or recommendations expressed in this material are ours and do not necessarily reflect the views of the National Science Foundation.

ABOUT THE AUTHORS

Erin Marie Furtak

Erin Marie Furtak is an associate professor of curriculum and instruction in the School of Education at the University of Colorado Boulder and specializes in science education. Her research grew out of her own experiences as a public high school teacher struggling to enact science teaching reforms in her own classroom. Her work focuses on how to support secondary science teachers in improving their everyday formative assessment practices.

Her research has been supported by the National Science Foundation, the Spencer Foundation, and the Alexander von Humboldt Foundation. She has been honored with the Presidential Early Career Award for Scientists and Engineers (2011) and the German Chancellor Fellowship from the Alexander von Humboldt Foundation (2006). Erin has published multiple peer-reviewed articles and two books, including one on the process of formative assessment design for secondary science teachers (published by Corwin in 2009). She provides extensive service to the teaching profession through long-term research and professional development partnerships with school districts and organizations in Colorado and across the United States.

Erin holds a BA in environmental, population, and organismic biology from the University of Colorado Boulder, an MA in education from the University of Denver, and a PhD in curriculum and teacher education from Stanford University.

Howard M. Glasser

Howard M. Glasser is a program officer of teacher development with the Knowles Science Teaching Foundation. His work has primarily focused on social justice and equity issues in education. He has also done extensive work around how educational technology affects teaching and learning and the importance of context, culture, and identity in influencing educational experiences and outcomes.

Howard has been a high school physics teacher in Philadelphia, taught preservice secondary math and science teachers, worked as a research associate for Research for Better Schools, and facilitated professional development activities for groups such as the Philadelphia Education Fund and the National Alliance for Partnerships in Equity. He has published peer-reviewed articles for practitioners and researchers in multiple journals. These papers have focused on a range of topics including how an inverted science curriculum affected student outcomes in science and math and how argumentation practices affected students' experiences in science classes.

He has a BA in physics with a concentration in educational studies from Haverford College; an MEd through Temple University's curriculum, instruction, and technology program; and a PhD in educational psychology and educational technology from Michigan State University.

ABOUT THE AUTHORS

Zora M. Wolfe

Zora M. Wolfe is an assistant professor of education at Widener University in Chester, Pennsylvania, where she primarily teaches in the K–12 Educational Leadership program. Her areas of expertise include teacher leadership, collaborative inquiry communities, and developing teacher practice. Her current research focuses on how principals can support teachers in their development as teacher leaders.

Zora's career has included a variety of educational experiences, from teaching kindergarten in an American school in Taiwan to teaching high school math and science in New York City and Denver, Colorado. She was part of the founding staff and principal of a charter high school and also has experience as an assistant principal and curriculum director at the K–12 levels. More recently, she worked with beginning math and science teachers across the United States, supporting their development as teacher leaders through an educational nonprofit organization.

She has a BS in psychobiology from Binghamton University; an MA in secondary science from Teachers College, Columbia University; and an EdD in educational leadership from the University of Pennsylvania.

CONTRIBUTORS

Stephan Graham
Science Teacher
*Arrupe Jesuit High School
Denver, Colorado*

Angela Haydel DeBarger
Program Officer
*George Lucas Educational
Research Foundation
San Rafael, California*

Christopher Harris
Senior Researcher
*Center for Technology in Learning
SRI International
Menlo Park, California*

Kate Henson
Science Teacher
*Miss Porter's School
Farmington, Connecticut*

Sara C. Heredia
Postdoctoral Researcher
*The Exploratorium
San Francisco, California*

Kelly Lubkeman
Chemistry Teacher
*Longmont High School
Longmont, Colorado*

Deb Morrison
Science Teacher
*Broomfield Heights Middle School
Broomfield, Colorado*

William Penuel
Professor of Educational Psychology
and Learning Sciences
*School of Education,
University of Colorado Boulder
Boulder, Colorado*

Katie Van Horne
Postdoctoral Researcher
*School of Education,
University of Colorado Boulder
Boulder, Colorado*

Erin Zekis
Physics and Math Teacher
*Arrupe Jesuit High School
Denver, Colorado*

INTRODUCTION

When we reflect on our reasons for writing this book, each of us can see how the seeds of the ideas it contains were sown many years ago. As full-time science educators, we loved teaching science and working with students but found ourselves struggling to find ways to efficiently explore our instruction in ways that could help us grow in our practice and improve student learning. We wanted to understand how different aspects of our teaching—whether they were new labs we created, different structures for group work, questions we wrote for tests, or problems we posed during lessons—influenced students and their learning. At the same time, we had difficulty figuring out how to select and use the information available to investigate our practice. We wanted to become better teachers but were overwhelmed with other tasks and uncertain how to begin examining these areas. Keeping up with planning and grading kept us incredibly busy and trying to actively generate and analyze other pieces of information was exhausting! We knew this work could be valuable but didn't know where to begin or how to spend our time wisely so the work would be feasible and useful.

Although we each continued to explore ways to improve our science teaching, we also began looking at how other teachers could enhance theirs, too. These interests led our paths to cross in 2011, when we developed and refined professional development experiences for novice teachers. We created resources and provided support that focused on aiding teachers in using data to improve their teaching and enhance students' learning. We have subsequently refined these materials and approaches and have used them with many groups of middle and high school teachers—not only those just getting started in their careers but also with those who have been in the profession for many years. Hearing these teachers' feedback further convinced us that these ideas could be valuable to more than just the specific groups with which we've worked. We wanted to write this book to share and discuss these ideas with more secondary science teachers.

In many schools today, there is a great emphasis for teachers to use "data-driven" approaches to teaching, and many teachers encounter pressure to use practices that will increase student scores on standardized tests. Often, these scores are the major pieces of data other people use to assess the effectiveness of teachers. This book broadens that perspective, drawing on approaches to assessment design (e.g., Atkin and Coffey 2003; Ayala et al. 2002; Pellegrino, Chudowsky, and Glaser 2001) and focusing on how data collected about student learning can help teachers improve their teaching and students' learning. To make this work more manageable, this book also provides numerous tools and resources that we developed in our collaborative work together, as well as in related research studies (e.g., Furtak and Heredia 2014; Furtak, Morrison, and Kroog 2014).

We build out from this focus on data to introduce a framework that we call the *Feedback Loop* as a model for how you can use data to explore your teaching, improve your practice,

INTRODUCTION

and enhance students' learning. Although this framework and the related ideas can be valuable for teachers of a variety of grade levels and subjects, the examples we provide focus on middle and high school science. Chapter 1 introduces the Feedback Loop, explaining its structure and usefulness in guiding teachers to consider a number of components when setting goals, developing tools to collect data, and analyzing those data to determine next steps for instruction. Chapters 2–5 each highlight one of the four elements: goals, tools, data, and inferences. We discuss the element's value, explaining what it is and how teachers can use it to grow in their practice and better support student learning.

Chapter 6 explores how to close the loop by connecting inferences and goals through feedback, and Chapter 7 uses the full Feedback Loop to describe an approach for planning and informing instruction. Finally, Chapter 8 discusses how to collaborate with colleagues when considering data and the Feedback Loop to further increase the effect that this work can have. Finally, Chapter 9 provides additional resources to explore.

Throughout the book, we build on the standards that teachers are expected to meet, with a particular focus on the *Next Generation Science Standards* (*NGSS*; NGSS Lead States 2013). We took this approach knowing that not all states have adopted the *NGSS* at the time of printing; however, since the states that have not each have their own form of standards, we took the *NGSS* because they are widely available and we could relate the classroom activities this book includes to them. Furthermore, we recognize that the *NGSS* represent the field's best knowledge of how science concepts and practices can be intertwined during instruction (NRC 2012), and as such, we feel that they are the best available source on which to base the examples. That said, many of the examples were developed pre-*NGSS*; several others took place afterward, in states that had their own sets of standards. While we related these ideas to the *NGSS* to bring them into the present science education policy context, we strongly emphasize that the Feedback Loop can work with any standards, be they *NGSS*, state, district, or even local school curriculum frameworks.

We have incorporated classroom vignettes in each chapter, as well as quotes and perspectives from current science teachers, to ground these ideas in real-life situations (note that all names marked with an asterisk are pseudonyms). We want this book to provide you with useful tools, approaches, and resources that will help you in your efforts to become better teachers. We hope you find these ideas and approaches as exciting and useful as we do!

References

Atkin, J. M., and J. Coffey. 2003. *Everyday assessment in the science classroom.* Arlington, VA: NSTA Press.

Ayala, C. C., Y. Yin, R. J. Shavelson, and J. Vanides. 2002. *Investigating the cognitive validity of science performance assessment with think alouds: Technical aspects.* New Orleans, LA: American Educational Research Association.

Furtak, E. M., and S. C. Heredia. 2014. Exploring the influence of learning progressions in two teacher communities. *Journal of Research in Science Teaching* 51 (8): 982–1020.

INTRODUCTION

Furtak, E. M., D. L. Morrison, and H. Kroog. 2014. Investigating the link between learning progressions and classroom assessment. *Science Education* 98 (4): 640–673.

National Research Council (NRC). 2012. *A framework for K–12 science education: Practices, crosscutting concepts, and core ideas.* Washington, DC: National Academies Press.

NGSS Lead States. 2013. *Next Generation Science Standards: For states, by states.* Washington, DC: National Academies Press. *www.nextgenscience.org/next-generation-science-standards*.

Pellegrino, J. W., N. Chudowsky, and R. Glaser. 2001. *Knowing what students know: The science and design of educational assessment.* Washington, DC: National Academies Press.

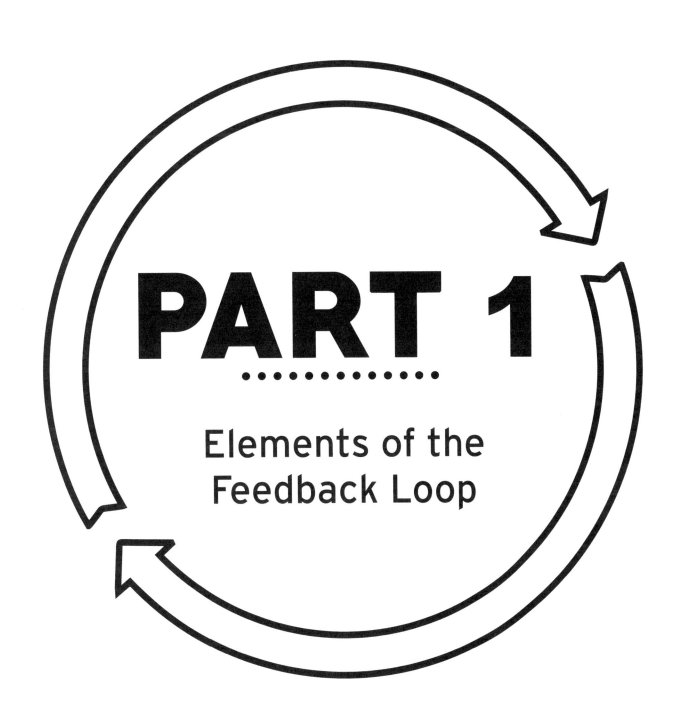

CHAPTER 1

Overview of the Feedback Loop

I use a lot of formative assessments, but I feel like I don't really appropriately analyze the data that I gather or use it for reteaching purposes in the most effective way.

—Beginning high school science teacher

Science teachers today are subjected to a deluge of data in their daily work. They are expected to draw on these data to make decisions about what to do during instruction. This trend, often called *data-driven decision making*, is the subject of big policy initiatives in schools right now. The amount of data coming at teachers *is* overwhelming. Let's take the case of an average public school secondary science teacher, who might be teaching upward of 150 students in five different classes a day. Although all of those classes might be repeats of the same course (e.g., eighth-grade physical science), it's more likely that the teacher has at least two different types of courses to teach. Given the current accountability system, it's likely that the students participate in standardized tests one or more times a year, yielding some kind of information for the teacher to interpret during or at the end of the school year. On a more immediate level, all of these students will be completing assignments and assessments daily or weekly. Although data is often confined to some type of written response, valuable data about student learning comes in many other formats, both formal and informal. Student comments during classroom discussions, the written work they generate—from lab reports to verbal responses to worksheets—and even student expressions and emotions can be considered data and can fill in gaps about what students know and are able to do in ways that test scores miss.

CHAPTER 1

To speak even more broadly, the preceding examples are all focused on students. You must also consider what you are learning about your teaching. For example, did the lesson go as you had planned? If students were doing a lab, was there a part that didn't come out as you expected? Did students take longer to finish than you thought they would? Was there a critical reagent that didn't work? Did you have to go from table to table telling students about a minor correction that was vital to the activity? When the lesson was over, did you think, "Well, I'm never doing it that way again," or "Next time, I'll change this part of the activity"? Interpreting and acting on these pieces of information, collected as you teach, is vital to adapting and improving over time and to helping students meet goals.

Taking a cue from the language of science, we might think about all these types of data as "noise" or some type of unwanted signal that can feel overwhelming and might seem easier to ignore. Furthermore, our training as science teachers can push against how we view these data, making us feel that they are not "controlled" or systematically collected and thus are unreliable. However, the idea of informing our daily decisions about instruction with data about what students know and are able to do is founded on the principle that there is a lot of "signal" mixed in with that noise, and it's the teacher's responsibility to find ways to sort through all the noise to identify the signal that they can use. It is important to look beyond the data you have and consider the specific goals you had in collecting it. For example, you may be interested in finding out how well the students understand a disciplinary core idea, such as Motion and Stability: Forces and Interactions, or you might want to assess how well the students are able to engage in a science practice, such as Analyzing and Interpreting Data.

All of these are different goals for which a teacher may want to collect data, and because they are distinct, the way the teacher engages in that data collection will likely vary. That is, to gather informative data that will actually help you assess your specific goal, you will need to use an appropriate tool.

The toughest thing about all information you must analyze and respond to as a science teacher—whether it's about what students are learning or doing or the activities that you give them—is that you must determine what it means. That is, you must look at a combination of students' work, statements in class, and interactions with a given activity, along with your supporting instructional approaches, to decide what students know and are able to do, what they are struggling with, and what that means for your instruction. Should you move ahead even though about half the students are still confused? Were the problems you encountered in the activity mainly a result of the way you introduced it, or were they due to the way the activity itself was written? If the students seem confused, what are they confused about, and are they all confused in the same way?

These questions all revolve around issues related to classroom assessment, that is, assessment conducted by teachers to ascertain what students know and are able to do. More specifically, these questions relate to what is often called *formative assessment*, or assessment conducted during the course of instruction for the purpose of fine-tuning to move students forward in their learning (Shepard 2000). This is distinct from *summative assessment*, which is usually administered at the end of a unit or the end of the year to determine what students know and assign them grades. The distinction between these two types hinges on the way that information about what students know and are able to do is used; if the assessment is conducted *on* learning, it is summative. In contrast, if the assessment is conducted *for* learning, or to improve or enhance the quality of teaching and learning in the classroom, it is formative (Wiliam 2007).

We refer to this distinction in classroom assessment because it relates very closely to the sets of questions raised about all the data produced about teaching and learning in classrooms. Reframing the way we think about those data helps us focus on what is most important and determine what those data are telling us about what students know, what they can do, and where they need help to move forward.

That's where this book comes in. Through our work with practicing secondary science teachers in designing and interpreting assessments and conducting inquiries into their own teaching, we have developed an approach designed to help you efficiently and systematically sort through all of this noise, extract meaningful information, and determine next steps for your teaching and for students' learning. In this chapter, we will zoom out from looking at data alone and present a framework for thinking about the data you are collecting, the goals you have for instruction, the activities you are using to determine what students know, and how you are deciding what students know on the basis of these elements. We will use a framework we call the Feedback Loop (see Figure 1.1, p. 8), which is intended to go beyond thinking about pieces of data in isolation to reorienting them as a part of a larger system that you, the teacher, can design and act on. While the processes in the Feedback Loop could be appropriate for teachers of any grade level and subject, in this book we focus on its specific applications to secondary science teachers and the unique ways its elements can guide our teaching and support student learning.

The Feedback Loop

Our framework for interpreting data is inspired by the methods that researchers and professional assessment developers have been using for years to design approaches for determining what students know and are able to do. Assessment developers never think just about the data they are collecting; instead, they develop assessments as part of a coherent process in which they consider what they want to assess, how they will assess it, what format the data will come in, and how they will interpret the data. There have been a number of ways that these processes have been described; for example, the National Research Council has an assessment triangle (Pellegrino, Chudowsky, and Glaser 2001), the Stanford Education Research Laboratory has an assessment square (Ayala et al. 2002; Ruiz-Primo et al. 2001), and the Berkeley Educational Assessment Research group has the BEAR assessment system (Wilson 2005).

CHAPTER 1

No matter what you call it or what shape it's in, all of these systems have four elements in common. To start, they all build on some form of the question, "What is the goal?" That is, rather than just looking at data, these frameworks deliberately focus on assessing something specific. For a classroom teacher, the goal should be the guiding principle that underlies what you are asking students to do. If students are doing a science lab, what is the reason you are having them do it? Are you interested in them coming to know a particular science concept, or are you interested in seeing the kinds of science practices in which they engage? For example, if you ask students to take measurements of water quality, such as pH levels and concentrations of dissolved substances in a local stream, is there a science practice goal, such as Obtaining, Evaluating, and Communicating Information, or are you more interested in supporting their learning of the crosscutting concept of Energy and Matter: Flows, Cycles, and Conservation?

This first step—being cognizant of the *goal*, or what you want to explore about your students' learning or your teaching—is the cornerstone of the Feedback Loop. Fortunately, policymakers have dedicated a lot of time working out what these goals are in the form of state or district standards or, more recently, the *Next Generation Science Standards* (*NGSS*; NGSS Lead States 2013), which define the disciplinary core ideas, science and engineering practices, and crosscutting concepts for students across grade bands. At the same time, you might have other goals that interest you about your own teaching that you wish to explore or strengthen, such as creating learning environments in which more students are able to participate in scientific argumentation or supporting students from diverse linguistic backgrounds in engaging in the language of science (e.g., Zembal-Saul, McNeill, and Hershberger 2012). These goals can also work as the starting point for the Feedback Loop.

The second element common among assessment development approaches is considering the answer to the question, "How will I know if students have met the goal?" The phrasing of this question leads to considering what we call the *tool*, or the activity, protocol, or other "thing" you are going to use to guide you in finding out what students know and are able to do. We use the term *tool* in the sense of an instrument that is used for a particular function, such as a meterstick to measure length or a spectrophotometer to measure wavelengths of light. In this case, we use a tool to find out what students know. In these examples, the meterstick and spectrophotometers are tools that help us to collect data in the form of different types of measurements. In the Feedback Loop, the tools are the common instruments teachers might use to collect data about student learning, such as worksheets, classroom assessments, external tests, and quizzes, as well as the questions a teacher might write to frame a classroom conversation, an observation protocol used to write down student ideas overheard from small groups, or even a tablet or smartphone used to record a lesson. It can also be something that is not written down or handed out but that you plan to use to get students to share their ideas, such as a really good, open-ended question asked as students engage in a laboratory investigation. The important common feature of all these tools is that

Overview of the Feedback Loop

they should be aligned with the goal you intend to assess. For example, just as scientists would not use a microscope to observe avian interactions at a distance, teachers should not use tools that are not aligned with the type of data they want to collect.

The third element of the Feedback Loop is *data*, a major focus of this book. Data are all the bits of information that can indicate what students know and that are yielded by the tools we use or create in our classrooms. Picking up from the examples above, the meterstick will yield data in the form of measurements of length, and the spectrophotometer will provide data in the form of measurements of wavelengths of light. In a classroom, data might be students' written responses to worksheets or classroom assessments, a teacher's written notes about student ideas shared in a whole-class conversation or small-group work, or the students' verbal responses to that open-ended question we mentioned above. Although data might seem very formal and official (think standardized test scores), they can also be unrecorded and ephemeral, such as the looks on students' faces when the teacher asks a particular question or a tally of student responses on a sticky note to help track student participation.

The last—and arguably most important—element of the assessment development frameworks has to do with how you make sense of the data you've collected. Since these elements are all connected, the process of making sense of the data is necessarily interwoven with the goals that you had and the tools you used. We call this process making *inferences* about what students know; that is, you're taking the multiple pieces of data you have and trying to determine what they tell you about the goals. When you think about it, it's very similar to the process of reasoning a scientist goes through. The individual pieces of information themselves don't necessarily make a whole lot of sense unless we consider them in light of why we collected them (goal) and what instruments we used to collect the data (tools). But if we consider these three elements together, we can piece together an argument about what the data might mean. For instance, you might infer from low standardized test scores that students have not met expectations for their grade band, or by watching a video, you might see that multiple students participated in a whole-class argument about evidence.

What is important about highlighting these four elements is that they should all be in sync with each other, and it is difficult to think about one element separately from the rest. We do not intend to suggest that the components should always be considered in a stepwise fashion; indeed, teachers often simply receive standardized test score data and must make sense of it. However, reflecting on the data along with the other three elements of the loop, rather than as individual pieces in isolation, empowers you to take ownership of what was being assessed in the first place, the nature of the tool that was used, and the inferences you can make from the data in hand.

Finally, after making inferences about what students know, the last step is to determine the implications for your teaching and supporting student learning relative to your learning goal. What have you learned about what students know and are able to do relative to the

CHAPTER 1

original goal? If students have not yet met the goal, what subsequent activities might you engage them in to help them move closer to it?

We bring all of these elements together in the Feedback Loop, shown in Figure 1.1. We represent them this way because, as we will argue in this book, thinking about one element in isolation misses the bigger context in which they are situated. For example, we find it impossible to think about data without thinking about why it was collected and the tool used to do so. Similarly, we find it difficult to make inferences about the implications any set of data might have about teaching and learning without considering why and how the data were collected.

FIGURE 1.1 The Feedback Loop

The Feedback Loop in Action: Uniform and Nonuniform Motion

To illustrate the different elements of the Feedback Loop, we will draw on one of our experiences in teaching middle school physical science. Erin was working with a group of seventh graders to find out what they had learned after a series of activities about measuring and representing uniform and nonuniform motion (those who have read Erin's 2009 book on formative assessment will be familiar with this example). Her *goal* was to have students interpret how strobe diagrams represent both uniform and nonuniform motion. At the same time, Erin wanted students to be able to construct arguments in support of their claims about what they observed, that is, to engage in the science practice of Engaging in Argument From Evidence.

Next, Erin went looking for a *tool* that she might use to determine if students were able to interpret strobe diagrams and make inferences about the speed of balls when given a representation of distance covered in constant amounts of time on two different ramps. After looking through several resource books such as *Physics by Inquiry* (McDermott et al. 1996) and some journal articles she had collected, she identified a set of three multiple-choice questions she intended to use to stimulate classroom discussion. Although multiple-choice questions are often thought of as being better fit for summative assessments, they can also work for formative purposes so that students sort themselves into categories in advance of a classroom discussion (Furtak 2009).

Once Erin had created the activity, she gave it to students and asked them to spend a few minutes individually answering the questions it contained. When all students had responded, Erin led a whole-class discussion in which students first voted on the answer they had selected and then provided reasons to justify their responses. In this way, the tool was designed not only to collect data about what students knew about representing motion, but also to determine their ability to engage in a science practice (Figure 1.2).

FIGURE 1.2 A tool created to assess student understanding of strobe diagrams and uniform and nonuniform motion

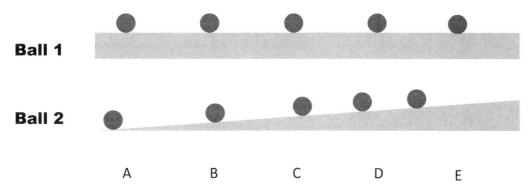

Source: Adapted with permission from Trowbridge and McDermott 1980.

1. How would you describe the motion of Ball 1?
    a. Uniform
    b. Nonuniform

How do you know?

2. How would you describe the motion of Ball 2?
    a. Uniform
    b. Nonuniform

How do you know?

3. At which point on the diagram do Balls 1 and 2 have the same velocity?
    a. Exactly at point A
    b. Exactly at point C

CHAPTER 1

c. Exactly at point B
d. Sometime between points B and C
e. Sometime between points C and D

How do you know?

The activity created two types of *data*: students' votes for each of the questions and their responses to the "How do you know?" follow-up questions. After asking students to vote on the answers to the first two questions, which asked students if they would describe the motion of each ball as uniform or nonuniform, Erin was able to ascertain quickly that the students understood the difference between the top diagram, which illustrated uniform motion, and the bottom diagram, which illustrated nonuniform motion (slowing down). She asked one or two students per question to share their reasons for selecting their answers and, noting the consensus of the class, moved to the third question.

Asking where in the diagram the two balls would have the same velocity generated more disagreement than the first two questions, with students providing arguments for response (c) and response (d). Erin made a number of *inferences* about what students knew on the basis of their clustering around these two responses. First, Erin knew from her prior teaching experience that a common student idea about uniform and nonuniform motion was that passing objects have the same velocity (Trowbridge and McDermott 1980). The division of students into two categories reflected that while about half the class was looking at the distance covered in a unit of time (response [d]), the rest of the class was identifying the point at which the balls passed as the point at which they had the same velocity (response [c]).

Finally, to take action on what she had learned about what the students knew, Erin decided to do the activity again and this time asked students to make repeated measurements of the ball on the ramp. Calling sets of students to the front of the room, Erin requested they take measurements and construct tables with sample data on both a flat surface and an angled ramp. She then invited students to calculate the speed between each of the strobe images, creating sets of speeds to compare. The students worked for several minutes and then quickly ascertained that the correct answer was response (d) (sometime between points B and C).

This example illustrates the way in which the four steps of the Feedback Loop guided Erin not only through the process of setting a goal, developing a tool, and collecting data, but also in interpreting those data and providing feedback to close the loop in advancing student learning.

Connections Between the Elements

It might seem that by zooming out and suggesting thinking about all of the elements of the Feedback Loop we are making the issue of interpreting data more, not less, complicated. It's true that we are asking you to think about more than just isolated pieces of data in

Overview of the Feedback Loop

making inferences about the quality of teaching and learning; however, we contend that thinking about the data along with the goals, tools, and inferences will ultimately help you feel *less* overwhelmed by the information collected. That is, rather than swimming in a pile of paperwork or feeling lost in the midst of a classroom conversation, you will go in one direction of the framework to consider how these data connect with goals. If the data are not aligned with goals, maybe you don't need to make inferences about it at all. Or, maybe your tools need to be adjusted so that the data you generate allow you to make better, more efficient inferences about what students know.

The Feedback Loop pushes us to consider the connections between the different elements, as shown in Figure 1.3. While a first step for a teacher might be to identify goals, tools, data, and inferences, the questions listed in the arrows in Figure 1.3 can help you evaluate the entire process of data collection and interpretation.

FIGURE 1.3 Connections between elements of the Feedback Loop

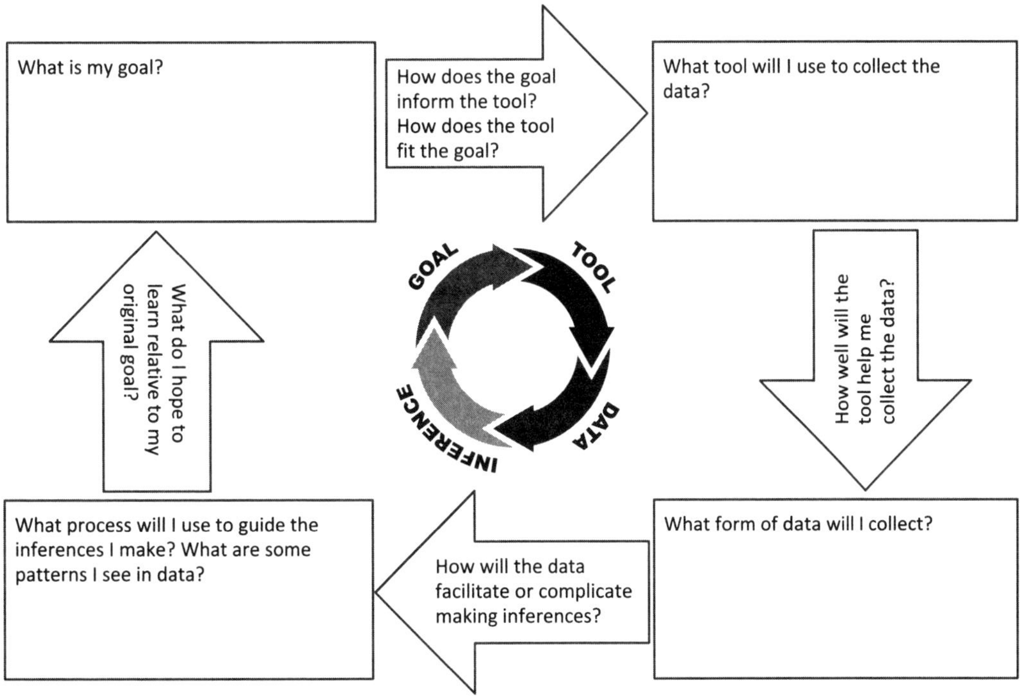

```
┌─────────────────────┐    How does the goal    ┌─────────────────────┐
│ What is my goal?    │──▶ inform the tool?  ──▶│ What tool will I    │
│                     │    How does the tool    │ use to collect the  │
│                     │    fit the goal?        │ data?               │
└─────────────────────┘                         └─────────────────────┘
         ▲                      GOAL  TOOL                 │
         │                    ╱           ╲                │ How well will
   What do I hope to         │   (loop)    │               │ the tool help me
   learn relative to my      │             │               │ collect the data?
   original goal?             ╲           ╱                ▼
                          INFERENCE  DATA
┌─────────────────────┐                         ┌─────────────────────┐
│ What process will I │◀── How will the data ──│ What form of data   │
│ use to guide the    │   facilitate or        │ will I collect?     │
│ inferences I make?  │   complicate           │                     │
│ What are some       │   making inferences?   │                     │
│ patterns I see in   │                        │                     │
│ data?               │                        │                     │
└─────────────────────┘                        └─────────────────────┘
```

The first connection between the goal and tool asks how a given goal informed the selection of a tool and how the tool fit the goal. The next connection raises the question of how well the tool helped you collect the data, and the third connection highlights how particular pieces or sets of data facilitated or complicated making inferences. Finally, the last connection between the inferences and goal help you evaluate what was learned with respect to the original goal that drove the process of data collection and interpretation.

CHAPTER 1

The Feedback Loop in Three Dimensions

However, we do not want to give the impression that the Feedback Loop exists in two dimensions. We see each feedback loop as building on other loops, and even growing upward, as you uncover new insights about what your students know or about your teaching and ask new questions that build on the ones that preceded them. Your inferences may help you determine your next goal; thus, you will design another tool and gather more data as your students gain understanding. Alternatively, your inferences may lead you to decide that you need more data or a different tool. By considering these elements as interconnected and iterative, teachers can continuously adapt and adjust their teaching to meet the needs of the learners in their classroom.

Figure 1.4 illustrates how we see multiple feedback loops building on each other over time. Figure 1.4a shows how inferences made on the basis of a feedback loop built on the initial goal might lead to another goal, and Figure 1.4b shows how each of these goals can lead to another feedback loop and ultimately another goal.

Looking Ahead

As the book continues, we will dedicate a chapter to each of the elements, providing rich examples and vignettes of practicing teachers working in different parts of the framework. We will also suggest resources and procedures for you as you explore each piece of the Feedback Loop as well as ways of engaging your colleagues in analyzing data using this approach.

Overview of the Feedback Loop

FIGURE 1.4 The Feedback Loop in three dimensions: (a) Inferences can lead to a new goal, and (b) these goals can lead to new feedback loops, which lead to new goals.

1.4a

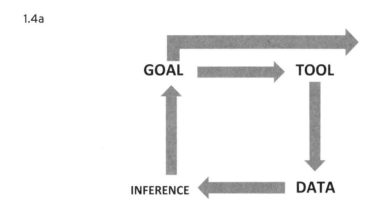

1.4b

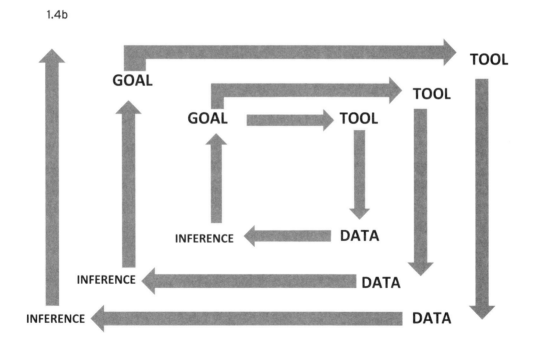

CHAPTER 1

References

Ayala, C. C., Y. Yin, R. J. Shavelson, and J. Vanides. 2002. Investigating the cognitive validity of science performance assessment with think alouds: Technical aspects. Paper presented at the annual meeting of the American Educational Research Association, New Orleans, LA.

Furtak, E. M. 2009. *Formative assessment for secondary science teachers.* Thousand Oaks, CA: Corwin Press.

McDermott, L. C., P. S. Shaffer, M. L. Rosenquist, and Physics Education Group at the University of Washington. 1996. *Physics by Inquiry.* Vol. 2. New York: Wiley & Sons.

NGSS Lead States. 2013. *Next generation science standards: For states, by states.* Washington, DC: National Academies Press. *www.nextgenscience.org/next-generation-science-standards.*

Pellegrino, J. W., N. Chudowsky, and R. Glaser. 2001. *Knowing what students know: The science and design of educational assessment.* Washington DC: National Academies Press.

Ruiz-Primo, M. A., R. J. Shavelson, M. Li, and S. E. Schultz. 2001. On the validity of cognitive interpretations of scores from alternative concept-mapping techniques. *Educational Assessment* 7 (2): 99–141.

Shepard, L. A. 2000. The role of assessment in a learning culture. *Educational Researcher* 29 (7): 4–14.

Trowbridge, D. E., and L. C. McDermott. 1980. Investigation of student understanding of the concept of velocity in one dimension. *American Journal of Physics* 48 (12): 1020–1028.

Wiliam, D. 2007. Keeping learning on track: Classroom assessment and the regulation of learning. In *Second handbook of mathematics teaching and learning*, ed. J. F. K. Lester, 1053–1098. Greenwich, CT: Information Age Publishing.

Wilson, M. 2005. *Constructing measures: An item response modeling approach.* Mahwah, NJ: Erlbaum.

Zembal-Saul, C. L., K. L. McNeill, K. Hershberger. 2012. *What's your evidence? Engaging K–5 children in constructing explanations in science.* Upper Saddle River, NJ: Pearson.

CHAPTER 2

Setting Goals

If the teacher has a clear road map that designates pivotal stops along the way, it is far easier to incorporate those stops.

—W. James Popham

Do you guys actually set goals before starting a unit? I would like to but am sadly at the point where I basically set a goal for the day.

—First-year high school teacher

Imagine setting out on a journey: The first step, of course, is determining where you are headed, and everything else follows from that. What will you pack? Will you need shorts or long underwear? Hiking boots or dancing shoes? What stops might you make along the way?

The next step might be to plot out your path. For much of human history, this involved using some kind of map to locate where you were starting and where you wanted to end up. Nowadays, this process is made much simpler for those with navigation systems or smartphones: Open an app or turn on your GPS, type in your goal, and the path is set for you. Either way, the route is laid out, and you just follow it to get to your destination.

Once you start on the journey, especially if it's a long one, you will probably stop to see how you're coming along. For example, Erin remembers long summer drives between upstate New York where her family lived and her grandparents' house in the Midwest, during which she counted mile markers along I-80 as her family rode in their Chevrolet Celebrity wagon across one state and then another: Ohio, Indiana, Illinois, and Iowa. The

CHAPTER 2

mile markers ticked off the progress they were making toward their ultimate destination: North Platte, Nebraska.

If you've ever been on a long car trip with kids, you can imagine how this 22-hour journey went. There were planned stopping points along the way (next stop: South Bend, Indiana! West Des Moines, Iowa!), but inevitably one of the three kids would ask the timeless question, *"Are we there yet?"*

This question gets at the fundamental nature of the Feedback Loop and the role that *goals* play in it. When you start out a car trip, you have a goal in mind. The Feedback Loop is the same way: It starts with identifying a goal for what you want your students to know and be able to do. Goals serve as the anchor for the Feedback Loop. They are the pins in the map that plot out your journey, both where you ultimately want to end up and your stops along the way. In this sense, they are a series of clearly articulated expectations. In the example of Erin's family vacation, the ultimate goal was North Platte, with intermediate goals of South Bend and West Des Moines. The sequencing of these goals helped Erin's parents to lay out the route they would take.

The question, "Are we there yet?" refers to an important function of learning goals: It defines the beginning of the process of the Feedback Loop that we will use to find out how students are making progress on their way toward the ultimate goal. Although the question could be taken rhetorically (as Erin's parents often treated it), it can also be authentic. An imagined response might be, "Well, we started off this morning in Latham, and we want to get to South Bend by evening time, but right now we're just passing Cleveland, so we have about two-hundred and fifty miles left to go." With the Feedback Loop, you will start with a goal, then compare data about students' progress to determine whether or not they have met that goal, and finally, use that information to identify next steps for teaching and learning.

Having clearly articulated goals is the foundation of the assessment design process (NRC 2001). It's essential to determine what it is that you want students to know and be able to do and to unpack what that will look like so you know it when you see it. Research has indicated that the clarity and coherence of teachers' goals for students are correlated with student achievement (Seidel, Rimmele, and Prenzel 2005).

In this chapter, we will get started by talking about how we think about goals in the Feedback Loop. First, we will talk about district and state standards as sources of goals in general, and then focus on how to unpack multidimensional standards such as the *Next Generation Science Standards* (*NGSS*; NGSS Lead States 2013). We will then provide a few illustrations of how these goals might be broken down before translating them into tools that will assess them (see Resource Activity 2.1, p. 34). Finally, we will talk about sequences of goals unpacked from the *NGSS*, or what some people call learning progressions, as ways of laying out those maps for where you and your students are headed (see Resource Activity 2.2, p. 35).

Setting Goals

Sources of Goals

The simplest incarnation of a goal for the Feedback Loop is a statement of your intentions for student learning in a given interval of time. Usually, this is clear declaration of what you would like students to know and be able to do; for example, you want students to be able to explain the difference between uniform and nonuniform motion, to describe a possible journey of an individual stone through the rock cycle, or to predict the combinations of alleles in the gametes of plants that are heterozygous for two traits.

What's important to note about these examples of goals is that they are both specific and measurable. That is, they are *specific* in that they state in detailed terms what the science is that students are expected to learn (i.e., the difference between uniform and nonuniform motion, allele combinations for heterozygous plants, or how an individual stone relates to the steps of the rock cycle), and they are *measurable*, in that they have directly observable performance of something students will be able to do ("explain," "describe," "predict").

These differ from the types of learning goals that we sometimes hear in secondary science classrooms, where a goal is often the equivalent of the activity students will do in a given day. When Erin was teaching, for example, a colleague told her that the day's goal was for students to do the "Lovely Liver Lab." Describing a goal in this way is neither specific (What science were students expected to learn and do?) nor measurable. Erin and her colleagues got together and looked closely at the lab, and it quickly became clear that the goals it embodied had to do with variables influencing enzyme activity, such as enzyme or substrate concentration, the presence of inhibitors, or temperature. Rather than "do the 'Lovely Liver Lab,'" the teacher's goal was to have students "collect and analyze data about the effect of temperature, enzyme and substrate concentration, and the presence of inhibitors on enzyme activity."

The difference between these two phrasings is key as you're jumping into using a feedback loop: the more specific the wording, the better positioned you'll be to select or design a tool to measure the goal, gather data related to it, and make inferences as to whether or not your students met it. Even stating the goal slightly more precisely, for example, "students will learn about enzymes," makes it neither specific nor measurable. In contrast, detailing exactly what aspects of enzyme activity students will learn about and then stating the goal in measurable terms by using the words "collect and analyze data" delineates exactly what students will be doing and makes it easier to translate the goal into a tool you will use to engage them.

When you're identifying a goal for the feedback loop, what you want to find out about what students know and are able to do may often be derived by digging into student activities in an exercise you're already doing; other times, you may need to identify the goal by unpacking the curriculum you're using.

CHAPTER 2

In many instances, your goals may be provided for you in the form of state or district standards, curriculum frameworks, or pacing guides in local use at your school. These will come in many forms, from the generic to specific. Regardless of the source, be sure that the goal you ultimately arrive at gets very specific about the science students are going to do and how you're going to know that they know and are able to do that science.

Sequences of Learning Goals

Even if goals are both specific and measurable, it feels a little artificial to talk about them in isolation. Middle school goals build on what students were expected to learn in elementary school. Curricula similarly build on a sequence of goals such that important preceding ideas are taught before those that rely on them. Every goal that is part of a feedback loop is embedded in sequences such as these, so it's useful to talk about how goals are related to each other before going too deep into the Feedback Loop method.

Lately, there has been a lot of talk about sequences of goals for students as learning progressions or representations of how student ideas and practices are expected to develop in a domain and over time (Duschl, Maeng, and Sezen 2011). Learning progressions are bordered on one end by less sophisticated concepts and the ideas that students have when they enter school and on the other end by the scientific explanations students are expected to learn. In the middle are the various intermediate understandings students may develop as they move toward competence (Corcoran, Mosher, and Rogat 2009; Alonzo and Gotwals 2012).

In the Feedback Loop, we can take up the idea of learning progressions as a way of mapping out a sequence of instruction and how ideas will build on each other. These sequences can help you to locate the places in your instruction at which it is most critical to stop and find out what students have learned so far. Erin has written previously about how thinking about a string of objectives can be represented as a staircase (Furtak 2009), starting with where you expect students to arrive in your class and climbing to the ultimate goal you want students to learn. Figure 2.1 shows a blank example of one of these staircase representations, and the vignette at the end of this chapter illustrates how teachers from a middle school created a staircase-style progression in a unit about the Earth, the Moon, and the Sun.

Unpacking the *NGSS* Into Goals and Progressions

So far, we've talked about goals and learning progressions in relation to different types of standards and curricula. We started from this more general approach since there are a lot of different ways that states and districts lay out the science students are expected to learn. However, since many states are now using the *NGSS*, we'll take them apart as a source of learning goals to illustrate how standards can be unpacked into goals for the Feedback Loop; however, if you work in a state that is not using the *NGSS*, keep in mind that you can adapt this process to work for any source of standards in use in your state, district, or school.

Setting Goals

FIGURE 2.1 Staircase learning progression

Source: Reprinted with permission from Furtak 2009.

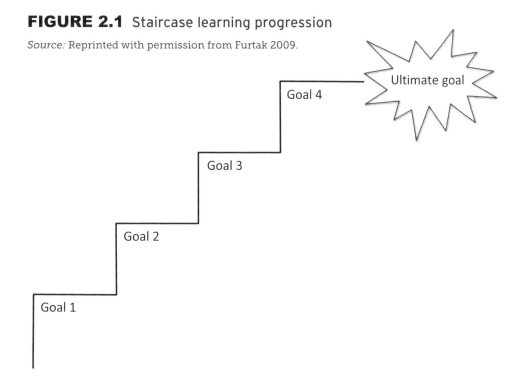

There are a number of big differences between the old *National Science Education Standards* (*NSES*; NRC 1996) and the *NGSS*. Both are built on decades of research into student learning and share the perspective that students should learn about the big ideas of science by engaging in the same activities that scientists do. However, the way that the standards are organized and written differs, as we'll illustrate below.

The *NSES* had two main components: a focus on inquiry-based science teaching and a list of content standards. The standards for inquiry were presented separately and listed what students were expected to do as they were learning the content standards: asking scientifically oriented questions, planning and conducting investigations, collecting data, using data to develop explanations, and communicating investigations and explanations. The content standards were presented next as statements of concepts students were expected to understand.

However, dividing up inquiry and content separately led to a difficult situation in which the emphasis was on "doing" science through hands-on investigations, and as a result, a wide range of activities of varying quality were fit under the umbrella of inquiry (Osborne 2014). Teachers interpreted the term *inquiry* in ways that were not necessarily consistent with what was intended by the *NSES* and sometimes misrepresented what scientists actually do (Windschitl 2004). Others noted that the idea of inquiry was often taught in a content-free manner, rather than intertwined with scientific ideas (Windschitl, Thompson, and Braaten 2008).

CHAPTER 2

As a result, the science education community set forth to define a new way of framing science standards that reflected a focus on practices (Osborne 2014), in what has been called the "practice turn" in educational circles, in which our focus is on engaging students in sets of practices rather than just content (Ford and Forman 2006). As a result, the *NGSS* barely discuss inquiry at all (the term inquiry appears 237 times in the *NSES*, but only 84 times in the framework for the *NGSS*) and instead focus on the science practices in which we want students to engage. These science practices look and sound a lot like the list above that broke inquiry down into smaller pieces; however, the major difference is that rather than being separated out into an inquiry standard, the science practices are woven into the content standards (Pratt 2013). At the same time, each standard includes crosscutting concepts, such as Patterns, Cause and Effect, and Structure and Function. Crosscutting concepts are intended to help students deepen both their understanding of disciplinary core ideas and the scientific and engineering practices.

Every standard in the *NGSS* is presented in a basic four-box structure that includes the title at the top followed by a list of performance expectations, and then three colored boxes containing the science practices, disciplinary core ideas, and crosscutting concepts that apply. We will give an overview of each process below; if you want more guidance for unpacking the *NGSS*, NSTA Press has created a number of resources that are listed in Chapter 9 for further reading. Figure 2.2 shows a complete four-box standard from the *NGSS*.

The advantage of performance expectations is that they make the process of translating standards into goals easier by already listing what students are able to do when they have met the standard. The challenge, then, is to set goals that get at all three dimensions of the *NGSS* at the same time. Disciplinary core ideas should not be taken out of the *NGSS*, and science practices should not be considered in isolation. Instead, assessments should be "designed to provide evidence of students' ability to use the practices, to apply their understanding of the crosscutting concepts, and to draw on their understanding of specific disciplinary ideas, all in the context of addressing specific problems" (NRC 2014, p. 32).

No one is denying that that's a tall order. Fortunately, the Feedback Loop can help you be sure that all three dimensions are pulled consistently through your process of designing tools, collecting data, and making inferences to guide your instruction. The first step is to unpack the three elements of an *NGSS* performance expectation to make them explicit. For example, according to MS-PS1-4, students should be able to "develop a model that predicts and describes changes in particle motion, temperature, and state of pure substance when thermal energy is added or removed." This performance expectation includes science practices (developing a model), crosscutting concepts (predicting and describing), and disciplinary core ideas (motion, temperature, and state of pure substance). Figure 2.3 highlights these same elements of the performance expectation for you. If you view the *NGSS* online, these elements will appear as pop-up boxes as you roll over the different parts of the expectation.

FIGURE 2.2 Sample *NGSS* box

Next Generation Science Standards: For States, By States

MS. **Structure and Properties of Matter**

PERFORMANCE EXPECTATIONS

Students who demonstrate understanding can:

MS-PS1-1. Develop models to describe the atomic composition of simple molecules and extended structures. [*Clarification Statement*: Emphasis is on developing models of molecules that vary in complexity. Examples of simple molecules could include ammonia and methanol. Examples of extended structures could include sodium chloride or diamonds. Examples of molecular-level models could include drawings, three-dimensional ball and stick structures, or computer representations showing different molecules with different types of atoms.] [*Assessment Boundary*: Assessment does not include valence electrons and bonding energy, discussing the ionic nature of sub-units of complex structures, or a complete depiction of all individual atoms in a complex molecule or extended structure.]

MS-PS1-3. Gather and make sense of information to describe that synthetic materials come from natural resources and impact society. [*Clarification Statement*: Emphasis is on natural resources that undergo a chemical process to form synthetic material. Examples of new materials could include new medicines, foods, and alternative fuels.] [*Assessment Boundary*: Assessment is limited to qualitative information.]

MS-PS1-4. Develop a model that predicts and describes changes in particle motion, temperature, and state of a pure substance when thermal energy is added or removed. [*Clarification Statement*: Emphasis is on qualitative molecular-level models of solids, liquids, and gases to show that adding or removing thermal energy increases or decreases kinetic energy of the particles until a change of state occurs. Examples of models could include drawings and diagrams. Examples of particles could include molecules or inert atoms. Examples of pure substances could include water, carbon dioxide, and helium.]

Science and Engineering Practices	Disciplinary Core Ideas	Crosscutting Concepts
Developing and Using Models Modeling in 6–8 builds on K–5 and progresses to developing, using, and revising models to describe, test, and predict more abstract phenomena and design systems. • Develop a model to predict and/or describe phenomena. (MS-PS1-1), (MS-PS1-4) **Obtaining, Evaluating, and Communicating Information** Obtaining, evaluating, and communicating information in 6–8 builds on K–5 and progresses to evaluating the merit and validity of ideas and methods. • Gather, read, and synthesize information from multiple appropriate sources and assess the credibility, accuracy, and possible bias of each publication and methods used and describe how they are supported or not supported by evidence. (MS-PS1-3) Connections to Nature of Science **Science Models, Laws, Mechanisms, and Theories Explain Natural Phenomena** • Laws are regularities or mathematical descriptions of natural phenomena. (MS-PS1-5)	**PS1.A: Structure and Properties of Matter** • Substances are made from different types of atoms, which combine with one another in various ways. Atoms form molecules that range in size from two to thousands of atoms. (MS-PS1-1) • Each pure substance has characteristic physical and chemical properties (for any bulk quantity under given conditions) that can be used to identify it. (MS-PS1-3) *(Note: This Disciplinary Core Idea is also addressed by MS-PS1-2.)* • Gases and liquids are made of molecules or inert atoms that are moving about relative to each other. (MS-PS1-4) • In a liquid, the molecules are constantly in contact with others; in a gas, they are widely spaced except when they happen to collide. In a solid, atoms are closely spaced and may vibrate in position but do not change relative locations. (MS-PS1-4) • Solids may be formed from molecules, or they may be extended structures with repeating sub-units (e.g., crystals). (MS-PS1-1) • The changes of state that occur with variations in temperature or pressure can be described and predicted using these models of matter. (MS-PS1-4)	**Cause and Effect** • Cause and effect relationships may be used to predict phenomena in natural or designed systems. (MS-PS1-4) **Scale, Proportion, and Quantity** • Time, space, and energy phenomena can be observed at various scales using models to study systems that are too large or too small. (MS-PS1-1) **Structure and Function** • Structures can be designed to serve particular functions by taking into account properties of different materials and how materials can be shaped and used. (MS-PS1-3) Connections to Engineering, Technology, and Applications of Science **Interdependence of Science, Engineering, and Technology** • Engineering advances have led to important discoveries in virtually every field of science, and scientific discoveries have led to the development of entire industries and engineered systems. (MS-PS1-3)

See connections to MS. Structure and Properties of Matter on page 305.

NEXT GENERATION SCIENCE STANDARDS — Arranged by Topics 211

MS. Structure and Properties of Matter

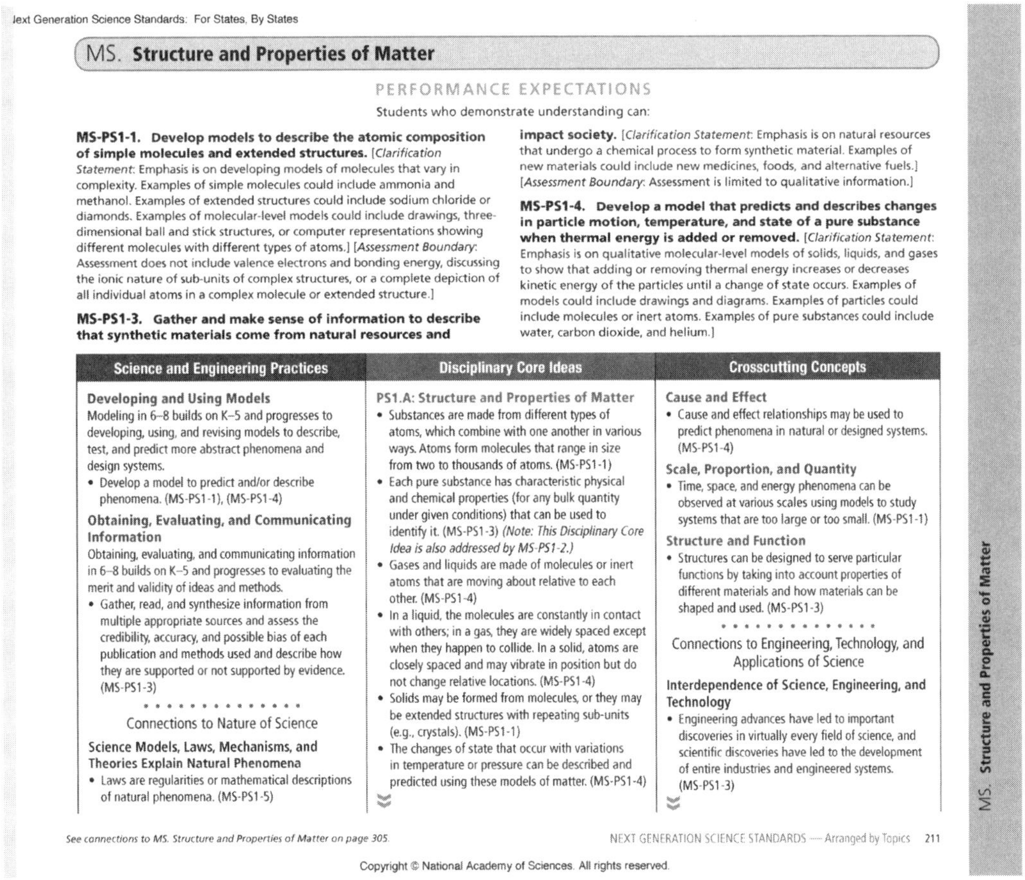

Source: NGSS Lead States 2013.

FIGURE 2.3 Sample standard with science and engineering practices, crosscutting concepts, and disciplinary core ideas labeled

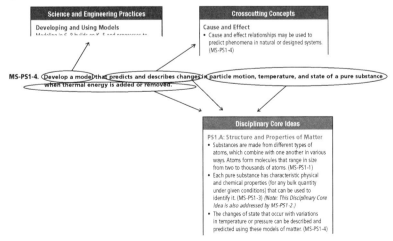

CHAPTER 2

As you get started in the Feedback Loop, we suggest you identify each of these dimensions in the standard, just to help you get your head around what the students are being asked to do. Resource Activity 2.1 (p. 34) provides a template for doing this unpacking, which will be essential to developing tools in the next chapter. The sample in Table 2.1 unpacks a standard for you.

TABLE 2.1 Unpacking *NGSS* MS-LS1-6

Standard or performance expectation	
MS-LS1-6: *Construct a scientific explanation based on evidence for the role of photosynthesis in the cycling of matter and flow of energy into and out of organisms.*	
***NGSS* DIMENSION**	**UNPACKING**
Science practice	Construct a scientific explanation based on evidence
Disciplinary core idea	Photosynthesis
Crosscutting concept	Energy and Matter

An important caveat to this process is that the connections among disciplinary core ideas, science practices, and crosscutting concepts in the *NGSS* are intended to be flexible. While each performance expectation includes all three of these components, you may determine that you want to combine them differently. The flexibility of the *NGSS* structure, as well as our process of unpacking the performance expectations, allows you to "mix and match" to better meet your own local learning goals.

The *NGSS* are built on learning progressions that represent how different performance expectations are related to each other, as well as what students are expected to be able to do at different grade bands. The great advantage of these is that they delineate specific goals as assessable performance expectations. At the same time, they lay these out in sequences intended to inform how teachers plan their instruction. However, it is important to note that

> *the target knowledge at a given grade level may well involve an incomplete or intermediate understanding of the topic or practice. Targeted intermediate understandings can help students build toward a more scientific understanding of a topic or practice, but they may not themselves be fully complete or correct. They are acceptable stepping stones on the pathways students travel between naïve conceptions and scientists' best current understandings. (NRC 2014, p. 37)*

In this way, the *NGSS* learning progressions can help to inform you not only of what your students' performance looks like for your end goal, but what intermediate steps on the way to that goal can look like.

Setting Goals

Exploring Student Thinking: Identifying Student Ideas as Resources

After identifying sets of goals for your students along with sequences or progressions of those goals, we recommend engaging in a crucial, final step: doing some research into how students learn the ideas and practices associated with that goal. We know from decades of research that students' ability to engage in scientific ideas is deeply intertwined with their previous impressions and lived experiences (NRC 2007). We also know that attempting to just "teach over" or "confront and replace" those ideas actually doesn't work; it just compels students to walk out of school without actually bringing their thoughts into conversation with each other.

A good example of this comes from Erin's experiences one Thanksgiving. As a graduate student, Erin worked with her colleagues to research a curriculum and associated assessments on students' conceptions about sinking and floating, some of which are included in this book. Erin spent literally every day for four years researching students' density-based explanations about what made things sink and float in water (e.g., students comparing the density of a sinking object to water would notice its density was greater than water, and vice versa for floating objects).

That year, Erin and her husband, Dave, decided to try frying a turkey. Dave poured water into the turkey frying pot and then set the turkey inside it to determine how much oil he would have to buy. Erin walked over and observed the turkey floating on top of the water, and wondered out loud how the turkey was going to be able to cook when it was floating and not sinking. Dave snickered, looked Erin in the eye, and said that it would not be a problem because he planned to cook the turkey in oil, not water, and since oil was less dense than water, the turkey would sink. "Aren't you supposed to be a sinking and floating expert, or something?" he asked, slyly.

Erin, embarrassed, realized she had grasped an important lesson about learning. She had spent a lot of time thinking about the right ideas, but she had not allowed those ideas to come into conversation with her prior experiences and everyday examples. As a result, she was able to simultaneously hold two explanations in her head—one based on her everyday life and another on the "science" explanation she had learned (and taught!) in school.

A key role of formative assessment is to surface exactly these kinds of conceptions. In the next chapter, we'll talk about designing tools that help get at students' everyday experiences. Even so, we've found that identifying what these impressions actually are as part of establishing learning goals is very helpful. The best resource we know that reviews literature on students' prior ideas is available at the National Science Digital Library (NSDL) science literacy maps (*http://strandmaps.dls.ucar.edu*). Each of these strand maps illustrates how key concepts within science and mathematics domains build on each other and how they are interrelated across strands. However, for each map, we also note that there is a small link

CHAPTER 2

in the upper left-hand corner that says "View Research on Student Learning." This link will take you to a listing, with citations, of students' prior ideas and experiences related to those in the map and will help you prepare to draw out those ideas as part of your trip through the feedback loop. Figure 2.4 shows the location of this link for the disciplinary core ideas represented in the life science standard we unpacked in Table 2.1.

This research into student learning lists several relevant ideas students may have about energy cycling and photosynthesis that are relevant to our learning goals in the feedback loop for MS-LS1-6 (see Table 2.1, p. 22, and Resource Activity 2.1, p. 34). Students tend to think that matter is created and destroyed but not transformed from one form to another. In addition, they know that cyclical processes take place in ecosystems but do not pay attention to the processes of plant growth or animals eating plants (*http://strandmaps.dls.ucar.edu/?id=SMS-MAP-9001*). We frame such student ideas as resources to be used in your planning of the assessment process rather than as misconceptions. This reflects our perspective that it is *good* to use the feedback loop as an opportunity to design tools that will surface these ideas, which you will then draw on them to inform your instruction. Resource Activity 2.1 at the end of this chapter includes a row to list these ideas as you move through the feedback loop.

Involving Students

Just as teachers have many ways to plan for and engage in aspects of the Feedback Loop, there are countless ways students can be integrated into its elements. As the teacher, you may decide how and when you want to expose your students to different pieces and for what purposes; we will provide sections on involving students in Chapters 2–6 as we discuss each element.

With respect to involving students in setting goals, students need to have an understanding of what is expected of them to monitor—and improve—the quality of their work (Black and Wiliam 1998). As Sadler (1989, p. 121) wrote, "The learner has to … possess a concept of the standard (or goal, or reference level) being aimed for" as a starting place for their engagement in the assessment process. Making these learning goals explicit to students at the beginning and end of class has been shown to be associated with increased learning (Seidel, Rimmele, and Prinzel 2005).

Among other purposes, sharing goals with students helps establish shared meaning for what counts as quality work (Coffey 2003). Looking at exemplars together of, for example, a well-articulated model or a well-structured explanation and then guiding students in a conversation about what makes these exemplars high quality can establish a common understanding within the class as to what students are expected to know and do.

Setting Goals

FIGURE 2.4 NSDL science literacy map with research on student learning

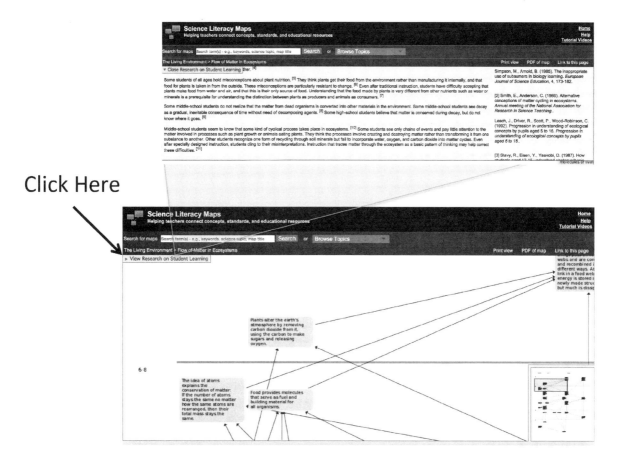

CHAPTER 2

A Staircase Progression in a Non-*NGSS* State

The administrators at Vasquez* Middle School asked their seventh-grade science team to develop common formative assessments for their astronomy unit and to use those assessments as ways to track student learning. The science teachers at Vasquez, like those at many schools in the United States, have several resources that they can pull from to help them plan assessments, including state standards and assessment frameworks, school district pacing guides, and the curriculum their school has adopted.

Vasquez is located in a non-*NGSS* state; however, its state standards integrate science practices with disciplinary core ideas. Its standards for astronomy in middle school note that students should model the relative positions of the Earth, the Moon, and the Sun and use this model to explain observable effects such as eclipses and Moon phases. Performance expectations for students in eighth grade note that, to score at the second-highest level of proficiency, students should be able to relate phenomena such as eclipses and lunar phases to rules governing the solar system.

The school district has placed learning about the Earth, the Moon, and the Sun at the eighth-grade level, and to support students in learning about their standards, it has adopted the Project-Based Inquiry Science (PBIS) curriculum, which is framed around Big Questions, which orient learners to the essential ideas in science, and Big Challenges, which guide instruction (IAT 2010). PBIS textbooks include sequences of instruction that build logically to help students answer the Big Questions described in each learning set.

After learning about the Feedback Loop in a district-sponsored professional development meeting, the science teachers at Vasquez gathered these resources to design a common formative assessment. They set their sights on a unit that was several weeks away in which student would engage with a series of models to support them in learning about the relative positions of the Earth, the Moon, and the Sun. The teachers would bolster this instruction with Learning Set 2 from PBIS-Astronomy, which engages students in a series of investigations of physical models of these three bodies.

---
* This and all other school and teacher names marked with an asterisk are pseudonyms.

Setting Goals

Vasquez's school district had worked with the "staircase" model of progressive learning goals for several years, so the teachers started by drawing a blank staircase with a star for the Big Question students would answer at the top. Although the overarching question driving the students' inquiry is, "How can you know if objects in space will collide?" Learning Set 2 is organized around the Big Question, "How do the Earth, Moon, and Sun move through space?" The teachers then revisited their state standard and related it to the subsections of Learning Set 2, extracting the major learning that students were expected to experience on the way to this ultimate goal.

First, the teachers jotted down the different types of models that students would build as they moved through the unit. Some sections were already phrased as questions, whereas others needed a little more unpacking. As the group read through, they noted not only progressive ideas for the staircase but also that for each subsection students were expected to build a model that represented the main idea from that section. They noted exactly what students were expected to model, and they then added to the staircase how this science practice supported student engagement in each of the goals listed.

2.1. What are the motions of the Earth, the Moon, and the Sun?

2.2. Model the apparent motion of the Sun.

2.3. Where is the Moon located, and how does it move?

2.4. Describe the motions of the Moon and the Earth.

2.5. What do eclipses tell you about distances to the Sun and the Moon?

The teachers also noted that for each of these investigations, the students would be constructing a model, and then they wrote down what the students would be modeling in each step.

Once they had mapped out these different learning goals, they cross-referenced the progression with the state standards and found that it mapped neatly onto what students were expected to know and do in middle school. They then filled in the staircase, beginning by placing the ultimate learning goal into the star at the top ("How do the Earth, the Moon, and the Sun move through space?") and then writing in what they expected students to do at each step, combining the questions from PBIS with the models students would create in the state standards (Figure 2.5, p. 28).

CHAPTER 2

FIGURE 2.5 Staircase learning progression with learning goals for astronomy unit

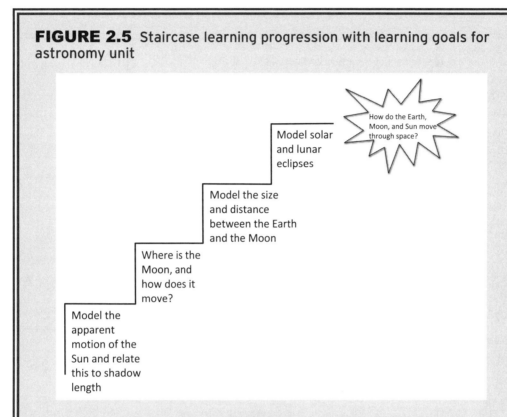

Source: Adapted with permission from Furtak 2009

The teachers had also learned that identifying students' common everyday ideas and misconceptions was an important starting place for designing formative assessments. For this information, they searched the NSDL science literacy maps (The Physical Setting → The Solar System → Phases of the Moon). After exploring the interrelationships among concepts in the map, they clicked the "View Research on Students Learning" box in the upper left-hand corner of the map. There, they learned that the idea that the Earth orbits the Sun can be counterintuitive to younger children, and students might not be able to understand how the Moon is illuminated by light from the Sun or how the phases of the Moon work before they understand the relative size and motion of the Earth, the Moon, and the Sun (Sadler 1987). They cross-referenced this information with their staircase and noted that its sequencing—modeling the relative size of the Earth and Moon before modeling the phases of the Moon and exploring sources of light and shadows—dovetailed with this research on student learning.

The group determined that they wanted to develop a common assessment that would provide opportunities for students to draw a model of the Earth, the Moon, and the Sun at the beginning and midway through their unit so that

> they could surface student thinking about the relative positions and motions of these objects in space. They then planned to look at the information from these assessments before students began to model phases of the Moon and to use information about it to support students.

Summary

This chapter talked about the importance of setting learning goals, walked you through features of the *NGSS* performance expectations and how to unpack them, and represented learning goals in staircase-style learning progressions. It also emphasized the importance of doing a little bit of research into students' common ideas as resources to inform the goal step of the Feedback Loop. In the next chapter, we'll take up these goals and move into how to develop them into tools.

CHAPTER 2

Quick-Survey Formative Assessment With Multiple-Choice Questions and Votes

Kelly Lubkeman, Longmont High School

My name is Kelly Lubkeman, and I am finishing up my first year teaching general and honors chemistry at Longmont High School. I work with an ethnically, linguistically, and socioeconomically diverse population, and I am continuously challenged to reach students of all ability levels and backgrounds.

I worked with Erin, who was one of my professors in my teacher education program, to identify a *goal* for the final unit of the year, stoichiometry. Although I'm not working in an *NGSS* state, we used them as a resource for our assessment planning. When Erin and I went to look at the *NGSS*, we were astonished to discover that the word "stoichiometry" is not in there! This led us to have a discussion about what stoichiometry is really about: the conservation of matter within a chemical reaction. When we framed the concept this way, it led us to HS-PS1-7: "Use mathematical representations to support the claim that atoms, and therefore mass, are conserved during a chemical reaction." I liked that this standard got at the big ideas behind stoichiometry and combined it with the science practice of using mathematical representations. This fit perfectly since the goal was to assess students' use of mathematics to support the law of conservation of matter.

Since I was hoping to gather quick, "on the spot" data about students answers, I decided to use Plickers (www.plickers.com; Plickers gives you a visual graph to show how many students are answering each answer for a multiple-choice question) and additionally create a space for students to provide reasoning to go along with their answers. The *tool* was adapted from a question that had four response options for students to indicate what molecules looked like before and after a chemical reaction, but we adapted it to name specific atoms (hydrogen and chlorine) and to have the images be accurate to the relative size of these atoms. I added two more questions and provided space for students to explain their reasoning. The tool is shown in Figure 2.6.

FIGURE 2.6 Ms. Lubkeman's molarity tool

1. The black circles represent hydrogen, and white circles represent chlorine. Which of the following could represent a chemical reaction between hydrogen and chlorine?

On your paper, explain why you chose the answer that you did.

2. Looking at answer choice D, if I start with one mole of hydrogen and one mole of chlorine, how many moles of HCl will be produced?
a. 1 mole
b. 2 moles
c. 3 moles
d. 0.5 moles

On your paper, explain why you chose the answer that you did.

3. If we start with 36 grams of chlorine (assuming we have enough hydrogen to react with all of it), how many grams of HCl will we end up with?
a. 1.02 g
b. 37.01 g
c. 18.51 g
d. 234.6 g

On your paper, show the mathematical work that helped you arrive at your answer.

I collected several forms of *data* with this tool. For some of my classes, I had my students answer on their own and explain their reasoning. I displayed the students' answers so we could all see them and told the students to talk to their neighbors and discuss their reasoning. If they wanted to change their answers at that point, they were welcome to. I did a screenshot of the Plickers data from each of the three multiple-choice questions in Figure 2.7 (p. 32). Although I ultimately collected the students' responses, I used their written work mostly so they would have a starting point for their discussions with their partners. I also flipped through the answers after class.

CHAPTER 2

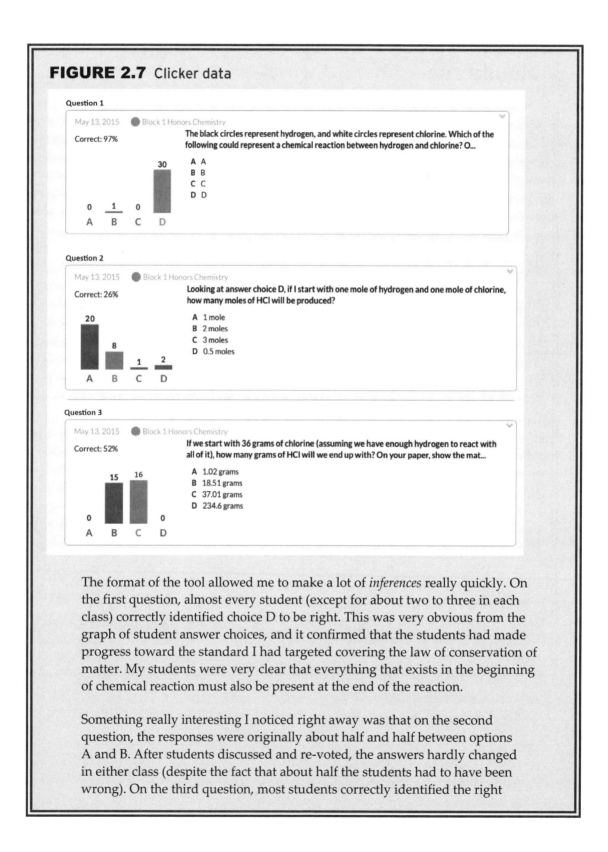

FIGURE 2.7 Clicker data

The format of the tool allowed me to make a lot of *inferences* really quickly. On the first question, almost every student (except for about two to three in each class) correctly identified choice D to be right. This was very obvious from the graph of student answer choices, and it confirmed that the students had made progress toward the standard I had targeted covering the law of conservation of matter. My students were very clear that everything that exists in the beginning of chemical reaction must also be present at the end of the reaction.

Something really interesting I noticed right away was that on the second question, the responses were originally about half and half between options A and B. After students discussed and re-voted, the answers hardly changed in either class (despite the fact that about half the students had to have been wrong). On the third question, most students correctly identified the right answer (choice C), but there were at least five or so in each class who didn't. Once the students answered and I saw what they chose, I had them discuss with their group members, and I saw a shift in answer choices to having almost every student choosing the correct answer. This allowed them to check with a neighbor to see where their mathematical mistakes occurred, and then they got to fix their mistake and re-answer. This question was really the piece that targeted our goal since we were hoping to assess student understanding about mathematical representations; they had to correctly use molar ratios to satisfy the law of conservation of matter.

I summarized my trip through the feedback loop in Figure 2.8. Plickers is such a nice tool to use because it gives instant feedback, and I can immediately see how students are changing their reasoning after they talk in small groups. Additionally, having the students write down their answers allowed me to go back later and determine what they were originally thinking when they picked their first answer. Ultimately, these data helped me see that although the students could track the atoms across the reaction in question 1, they were not immediately able to translate the reaction into moles or grams, as requested in questions 2 and 3.

FIGURE 2.8 Feedback loop for molarity assessment

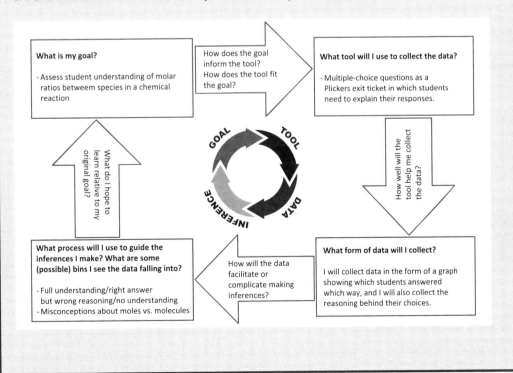

CHAPTER 2

RESOURCE ACTIVITY 2.1
Unpacking an *NGSS* (or Other Type of) Standard

Unpack the NGSS performance expectation you're working with into the science practice, disciplinary core idea, and crosscutting concept. Then, brainstorm how a multicomponent tool would draw out student thinking and practices for each element. If you are working with something besides the NGSS, substitute the language your district or state uses to describe the standard you are working with in the top box and the component pieces in the boxes below.

SAMPLE

Standard or performance expectation
MS-LS1-6: *Construct a scientific explanation based on evidence for the role of photosynthesis in the cycling of matter and flow of energy into and out of organisms.*

NGSS DIMENSION	UNPACKING
Science practice	Construct a scientific explanation based on evidence
Disciplinary core idea	Photosynthesis
Crosscutting concept	Energy and Matter
Student ideas as resources	Cyclical processes take place in ecosystems, but students do not attend to the processes of plant growth or animals eating plants. Students think that matter is created and destroyed rather than transformed (*nsdl.org*).

YOUR TURN

Standard or performance expectation	
NGSS DIMENSION	UNPACKING
Science practice	
Disciplinary core idea	
Crosscutting concept	
Student ideas as resources	

Setting Goals

RESOURCE ACTIVITY 2.2
Defining a Staircase Learning Progression

Use the staircase below as a place to jot down your ultimate learning goal and the stepwise learning goals you've set for your students.

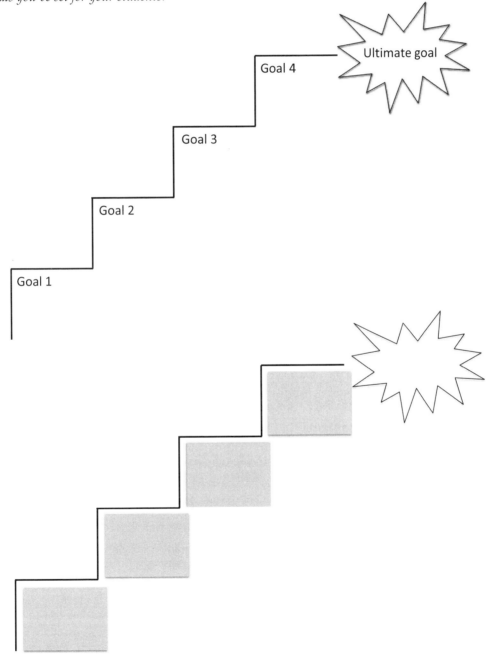

Source: Reprinted with permission from Furtak 2009.

CHAPTER 2

References

Alonzo, A. C., and A. W. Gotwals, eds. 2012. *Learning progressions in science: Current challenges and future directions.* Rotterdam, the Netherlands: Sense Publishing.

Black, P., and D. Wiliam. 1998. Inside the black box: Raising standards through classroom assessment. *Phi Delta Kappan* 80 (2): 139–148.

Coffey, J. 2003. Involving students in assessment. In *Everyday assessment in the science classroom*, ed. J. M. Atkin and J. E. Coffey, 75–87. Arlington, VA: NSTA Press.

Corcoran, T., F. A. Mosher, and A. Rogat. 2009. *Learning progressions in science: An evidence-based approach to reform.* Philadelphia, PA: Consortium for Policy Research in Education.

Duschl, R., S. Maeng, and A. Sezen. 2011. Learning progressions and teaching sequences: A review and analysis. *Studies in Science Education* 47 (2): 123–182.

Ford, M. J., and E. A. Forman. 2006. Chapter 1: Redefining disciplinary learning in classroom contexts. *Review of Research in Education* 30 (1): 1–32.

Furtak, E. M. 2009. *Formative assessment for secondary science teachers.* Thousand Oaks, CA: Corwin Press.

It's About Time (IAT). 2010. Project-based inquiry science. *www.iat.com/courses/middle-school-science/project-based-inquiry-science/?type=introduction*.

National Research Council (NRC). 1996. *National Science Education Standards.* Washington, DC: National Academies Press.

National Research Council (NRC). 2001. *Classroom assessment and the National Science Education Standards.* Washington, DC: National Academy Press.

National Research Council (NRC). 2007. *Taking science to school: Learning and teaching science in grades K–8.* Washington, DC: National Academies Press.

National Research Council (NRC). 2014. *Developing assessments for the Next Generation Science Standards.* Washington DC: National Academies Press.

NGSS Lead States. 2013. *Next Generation Science Standards: For states, by states.* Washington, DC: National Academies Press. *www.nextgenscience.org/next-generation-science-standards*.

Osborne, J. 2014. Teaching science practices: Meeting the challenge of change. *Journal of Science Teacher Education* 25: 177–196.

Pratt, H. 2013. *The NSTA reader's guide to the Next Generation Science Standards.* Arlington, VA: NSTA Press.

Sadler, D. R. 1989. Formative assessment and the design of instructional systems. *Instructional Science* 18: 119–144.

Sadler, P. 1987. Misconceptions in astronomy. In *Proceedings of the second international seminar misconceptions and educational strategies in science and mathematics.* Vol. 3, ed. J. Novak, 422–425. Ithaca, NY: Cornell University Department of Education.

Seidel, T., R. Rimmele, and M. Prenzel. 2005. Clarity and coherence of lesson goals as a scaffold for student learning. *Learning and Instruction* 15 (6): 539–556.

Windschitl, M. 2004. Folk theories of "inquiry:" How preservice teachers reproduce the discourse and practices of an atheoretical scientific method. *Journal of Research in Science Teaching* 41 (5): 481–512.

Windschitl, M., J. Thompson, and M. Braaten. 2008. Beyond the scientific method: Model-based inquiry as a new paradigm of preference for school science investigations. *Science Education* 92 (5): 941–967.

CHAPTER 3

Designing, Selecting, and Adapting Tools

It doesn't have to be something huge. It just needs to be a quick checkpoint and move on because it gives kids a chance to draw, sketch, describe. I think that's the key. We're already pretty overwhelmed with grading work and not having enough time for that. I need something that's going to work within my day with multiple lessons at a time for a checkpoint. Efficiency.

—Middle school science teacher

Imagine that Joe Rossman* and Alex Martin*, both middle school science teachers, are about to start a unit on kinetic energy with their eighth-grade physical science classes next week. They have identified one major goal, or performance expectation, that they'd like their students to meet. They want their students to "construct, use, and present arguments to support the claim that when the kinetic energy of an object changes, energy is transferred to or from the object (MS-PS3–2)." Although this goal explicitly uses language from the *Next Generation Science Standards* (*NGSS*; NGSS Lead States 2013), it's similar to goals they pursued before their state adopted these standards. Although they know that this is what they'd like their students to reach and they have a set of lessons to use, they need to find a way to assess the degree to which students are meeting this goal. Since the goal—like all the *NGSS* performance expectations—combines science practices along with content, they will need a tool that assesses both of those aspects. What advice would you give them?

The tools we use in our classrooms create opportunities to surface valuable information about what students know and are able to do. How you develop and sequence activities in a lesson,

CHAPTER 3

the instructions or information provided on a worksheet, and the questions you ask during a class discussion are just some of the many tools you can create and modify to generate rich data about student understanding. This chapter will walk you through processes for selecting, adapting, and designing tools to use in the Feedback Loop.

What Do We Mean by Tools?

The word *tool* is maybe not one you use in your daily teaching practice. More often, the word tool brings up images of a garden shed with rakes and shovels leaning against a wall or a workbench with screwdrivers, hammers, and pliers hanging from a pegboard above it. We like to use the word *tool* in its general meaning: a device or implement used to carry out a particular function. As science teachers, we spend our days surrounded by all sorts of instructional tools, such as microscopes, mass balances, test tubes, and Bunsen burners. In the context of the Feedback Loop, though, we are going to use the word *tool* in the very specific sense of some kind of classroom activity that creates opportunities for teachers and students to bring to the surface data about what students know and are able to do relative to a goal.

By using the word *tool*, we are able to be broad and take into account a wide range of teaching practices, instructional resources, and classroom activities. In fact, there are almost an infinite number of tools we could use to draw out student understandings and practices. Although this immeasurable number can be daunting, we think it's also incredibly exciting to acknowledge the many ways we can learn more about what students know and can do.

Tools are the currency of teaching; schools and classrooms are full of them (teacher guides to textbooks, online test banks, etc.). They are shared among teachers and schools at professional development meetings and national conferences. In fact, tools are so ubiquitous that Grossman, Smagorinsky, and Valencia (1999) have called the teaching profession "tool-centered." These tools can include worksheets, readings, laboratory handouts, rubrics, quizzes, tests, and exit tickets as well as questions teachers ask, instructional strategies to organize classroom talk, and graphics or models shared with students.

What Makes a Good Tool for the Feedback Loop?

Even though tools such as these surround us, not all are created equal, nor are they likely to be well aligned with your goals unless you engage in an intentional process of selecting and adapting existing tools or designing new ones. Tools that work well in the Feedback Loop will surface student ideas during instruction so you can use that information to further the students in their learning. That is, tools serve a formative function: They provide data aligned with your goal to guide your inferences about student thinking and learning (Black and Wiliam 1998).

This is different from the way we commonly think of tests or quizzes—also examples of tools—that we use primarily to assign grades. These types of tools, which can also be thought

Designing, Selecting, and Adapting Tools

of as *summative assessments,* are not as well-suited to the Feedback Loop because they take place on longer timelines that are necessary to score them carefully and assign grades. We want tools that help you make quick, informed decisions to guide your instruction and that are deeply grounded in data about what students know and are able to do.

For a tool to serve a formative function, though, it needs to adhere to a set of criteria that ensure it is designed and used to generate information. Formative tools are aligned with goals, focused on big ideas, and able to generate easily interpreted data that make student thinking visible.

Align the Tool With Goals

The first criterion for a high-quality formative assessment tool is that it be well aligned with a learning goal. If you want to know how students are able to use evidence generated from models to develop a scientific explanation for the causes of the seasons, your tool would need to have elements featuring all pieces of this performance expectation. These include using evidence from models, providing scientific explanations, cause-and-effect relationships, and the causes of the seasons. A tool that focused only on modeling, with no scientific content, or that tapped simple factual knowledge about the seasons would not be considered as aligned.

Without connecting your goal and tools, the latter are more likely to generate superfluous information that will not provide much insight into the area(s) that interest you the most. Although it might seem like that would only be a minor inconvenience, it makes the assessment less targeted to what interests you the most, creates additional work for students, pushes students to spend extra time in class or at home on things that are not as important to you, and can lead you to spend more time reviewing student work in areas that are not central to your main interests.

All of these outcomes pull on students' time as well as your own, resulting in less focus and attention on what matters most. To make your work in the Feedback Loop as efficient as possible, take the goals you so carefully developed and then thoughtfully select, adapt, or design tools that will support you in focusing on those goals. This takes a little more time up front but will ultimately enable you to develop more targeted tools that will save time for you once you start collecting data and making inferences—trust us!

Situate the Tool in a Problem Context

Good formative assessment tools don't waste time on the minutiae of science (Furtak 2009); rather, they are oriented around how the big ideas of science manifest themselves in everyday experiences or problem contexts, or what some in science education call "anchoring events" (AST 2014). These events are either observable directly or by instruments, and examples include earthquakes, water moving in and out of a cell, the diffusion of dye in a beaker, or

CHAPTER 3

the cooling of coffee in a mug. These events create opportunities for students to pull together what they know and use those experiences to explain the event (AST 2014). They all have underlying explanations that will get at the true nature of students' thinking about your goal, and they can be adapted into questions that will drive students to respond to the tool you are creating.

Picture it: What would be more compelling to a student, a dry learning goal written on the board or a fascinating question? Anchoring events cast performance expectations as compelling phenomena that students will be driven to explain and that connect sometimes abstract scientific ideas and practices with students' everyday experiences. They allow us to recast science learning as the answers to these arresting questions; for example, you could explain how a person can break a glass with their voice rather than teaching about sound waves or get students to understand how a bicycle left out in the rain can rust rather than lecturing on oxidation (AST 2014). The concepts and practices are, of course, assessed on tools designed around these intriguing questions, but the lead-in is something to which students can relate.

Generate Data to Make Student Thinking Visible

The next important feature of a high-quality formative assessment tool is that it is designed in a way that will make student thinking visible. This is a combination of both the type of question asked and the space provided for students to share their thoughts.

The first component of making student thinking visible is to ask questions that get at what students are really thinking. That means that instead of asking questions that start with "what"—that is, declarative questions (Li, Ruiz-Primo, and Shavelson 2006)—you focus on asking questions that get at *why* students think something might happen or *how* it might happen. These latter questions are more likely to get to the heart what is driving the anchoring events mentioned above. If you find that you can't meet a goal without asking a "what" question, we suggest that you add, "Why do you think so?"

The second component has to do with the space that you leave in the formative assessment tool for students to respond. Assessment designers talk about the *outcome space* (NRC 2014) of a tool to capture this quality and focus on whether an assessment has a constrained outcome space—think multiple-choice questions or fill in the blank—or an open outcome space, where students can really share their thinking. Providing students with an adequate amount of space cues them to expand and explain instead of simply providing a short yes or no answer.

Create Easily Interpretable Data

Students' written responses in a science notebook might very well make student thinking visible; however, if you have upward of 150 students and each student has written two to three pages in their notebook, this information is not in a format that is readily or easily interpretable. Creating time to use a multipart rubric to evaluate student responses, and even finding time

Designing, Selecting, and Adapting Tools

to read those notebooks, is unlikely to help you get the information rapidly enough to act on it in your classroom. You are much more likely to be able to use the data from the students to inform your teaching if you design a tool that generates information you can easily access.

Thus, the final criterion for a high-quality formative assessment tool for the Feedback Loop is that the information it generates facilitates easy interpretation. The best way to do this is to keep the tool short and simple. If it's an activity or series of questions, think about which questions will most quickly and easily yield the information you need. If you're asking students to draw a model, have them do it on a whiteboard so they can hold it up and you can see it quickly. If you're using a multiple-choice question, have them vote on the responses by raising hands or using clickers (if you have access to them). One of the easiest ways to design a tool to produce easy-to-access information is to have one good question that students can discuss in small groups and then either have them share their answers with the class or simply listen in on their conversations yourself.

We'll also note the importance of a process we call tool reduction. When considering the connections among your goals and tools, you might feel really confident about a tool you'd like to use in your class. However, it's also possible that you have identified *multiple tools* that could be good or a single tool that has multiple components. While this situation might not sound concerning, we have worked with teachers on the Feedback Loop who felt completely overwhelmed by the volume of data their tool generated. Although you can reduce the amount of data you've collected, it's also worth thinking about this when you're selecting, adapting, and designing tools so that the tool itself is shorter.

It can be helpful to think about how you'd select what tool(s) to use from among a set that all seem good. Pressure to teach other concepts and practices might limit the amount of time you have to focus on any goal, making prioritizing especially important. In the past, we've been excited about multiple tools, but even when we had time to use many of them, we found that doing so provided too much data to sift through effectively and typically led to redundant conclusions.

Selecting Tools

If you're like most teachers, you've already got a lot of resources around you that could be used as sources of tools. This is why we talk about selecting and adapting tools. For example, Page Keeley has written a number of books, listed in Chapter 9, that are filled with excellent formative assessment probes for multiple content areas and grade levels (e.g., Keeley, Eberle, and Farrin 2005). You may need to think more specifically about your goals to determine which probe might best help you get the information you need to assess your students' thinking in relation to your goal. In addition, you might need to ask yourself if the student responses that are elicited by the probe would actually give you information you want, or if you might want to adapt the probe to fit your goals.

CHAPTER 3

Resource Activity 3.1 (p. 65) provides places for you to unpack each of the three elements of *NGSS* performance expectations as well as space for you to brainstorm ideas for how you might assess each of these components. Once you've identified a candidate tool, we suggest you use Resource Activity 3.2 (p. 66) to evaluate its quality, not only with respect to the standard or *NGSS* performance expectation you've identified, but also in relation to the criteria for quality tools described earlier in this chapter. Resource Activity 3.3 (p. 67) provides space for you to brainstorm ideas you might hear from your students.

Adapting Tools

The arrival of the *NGSS* and its new framework for assessment (NRC 2014) means that many of the pre-existing tools that might have worked for the previous standards may need to be adapted to fit with the three dimensions. Tools that support the *NGSS* need to be *multicomponent*, or composed of multiple questions or activities that allow you to make inferences about whether students are able to engage in all three aspects of the *NGSS* (NRC 2014). These multicomponent tools might combine a space for students to draw a model, with room for writing that explains it (Kang, Thompson, and Windschitl 2014) or pair a carefully-selected multiple-choice question with a short-answer prompt (Furtak 2009). Older tools may just need an additional question, a small change in wording, or additional space for drawing to bring them up to the level of the new standards. Completing Resource Activity 3.2 will help you identify areas that your tool may need to be adapted or improved.

It's also possible that you may need to take an existing demonstration or activity and adapt it into a tool for the Feedback Loop. Take, as an example, the *NGSS* high school standard for Forces and Interactions, which states that students should "plan and conduct an investigation to provide evidence that an electric current can produce a magnetic field and that a changing magnetic field can produce an electric current" (HS-PS2-5). Assessment of this standard builds on middle school expectations of students and asks them specifically to test and provide evidence for different types of models. Simple factual recall questions that, in the past, may have asked students to define a magnetic field or even to describe how a changing magnetic field generates an electric current do not entirely match the standard. Instead, students must now additionally illustrate their ability to design and carry out an investigation that generates evidence in support of this relationship between an electric current and a magnetic field.

A common demonstration or classroom investigation to illustrate this relationship involves the teacher moving a magnet through a solenoid and measuring the current generated with a galvanometer. The *NGSS* expectation repositions this show-and-tell or experiment-and-confirm approach by asking students to engage in the activity themselves, perhaps by building a model of what they think is happening, testing it, and collecting data to evaluate it.

A formative assessment tool that will engage students in this complex practice-and-idea relationship, then, might be given after students had completed an iteration of developing, testing, and revising their model. A teacher might make small tweaks to the activity by, for example, inviting students to propose an investigation in which the number of coils on the solenoid was varied and the relationship of that number with the current generated was tested. To respond to such a question, students would need to simultaneously draw on their understanding of the association between electrical current and changing magnetic fields and their ability to design an experiment to test that association. The types of responses students provide would similarly help the teachers evaluate students' understanding of the disciplinary core idea (e.g., Do students know that the magnetic field needs to be *changing* to generate the current?) and their ability to design an experiment that would generate relevant data (e.g., How are dependent and independent variables controlled? What types of data is collected?).

Designing Formative Assessment Tools

If you're having trouble finding a tool that fits your goal, designing a new one might be an approach to consider. This can be tough, though, and if you try, we recommend you work with your colleagues to unpack standards and then use an existing assessment format to make this process easier.

We won't attempt to be exhaustive in suggesting formats here. For the purposes of simplicity, we will suggest two categories of formative assessment tools that surface student thinking: written and discussion. Written tools are stand-alone ways for individual students to surface their understandings, and discussion tools reveal student ideas in small- or whole-group discussions.

Written Tools

If you want students to create a written record of what they know and are able to do, written tools are a good place to start. Erin wrote about several formats for written tools in her 2009 book. We'll summarize some of them briefly here.

Evidence-to-explanation: This format for formative assessment provides students with some form of evidence, such as a figure, table, or graph, and asks them to provide an explanation for it.

Constructed response: This is basically a well-formulated question with a lot of space for students to write, draw models, or otherwise represent their thinking.

Multiple-choice plus justification: These types of questions are linked with students' common everyday ideas and allow students to select ideas and then explain why they chose them. Briggs and colleagues (2006) called these types of multiple-choice items distractor-driven; the responses that are not correct are linked with students' common misunderstandings and everyday ideas. This way, when students select one of these options, it provides you with more information rather than just whether they are right or wrong.

CHAPTER 3

The widespread use of personal-response systems, also called clickers, has made using multiple-choice questions in the classroom something teachers can do every day to generate easily interpretable information about student thinking. All they have to do is display a question, and then students can vote with the clickers and the response system aggregates the data and instantly displays it in a graph.

As an alternative, some teachers are using technologies such as Socrative (*www.socrative.com*), which allows students to enter responses on smartphones, tablets, or laptop computers, depending on which technologies are available. The system again aggregates or color codes the data for the teacher.

If you don't have access to such programs, there are other low-tech solutions that can get at the same idea. One option is to use colored pieces of paper that correspond with response options and have students hold them up to vote (or to just do a simple hand-raising). Other teachers use a "four corners" approach, in which they identify different corners of the classroom with the response options and ask students to stand in the corner that corresponds with their answer.

Predict-observe-explain: This type of tool, credited to White and Gunstone (1992), involves sharing a phenomenon with students and asking them to make a prediction of what they think will happen. Students next observe you demonstrating the phenomenon and then explain what they saw. A variation is the predict-explain-observe-explain, in which students also include an explanation for their prediction.

Models: The practice of creating and revising models to represent the mechanisms underlying scientific explanations is a big part of the *NGSS* reforms. Asking students to draw their models is another simple strategy. The Tools for Ambitious Science Teaching website (*http://ambitiousscienceteaching.org/tools-face-to-face*) contains a number of different forms of modeling tools that can be used as structures for students to record and share their models as a way of making their thinking explicit. For example, students can make summary tables that keep track of changes made to their models over time, or the teacher can direct students to make a consensus model that represents the thinking of the whole class or can provide students with a checklist to guide the ways in which they construct models.

Discussion Tools

In contrast with written tools, discussion tools are those that you don't necessarily print out or have students write responses to but that are questions or scenarios you pose to elicit ideas. There is a rich history, both inside and outside the formative assessment literature, of using classroom discussions as a way to surface student thinking (e.g., Duschl and Gitomer 1997; Michaels et al. 2013). Richard Duschl and Drew Gitomer (1997) have written about what they call *assessment conversations*, which are whole-class discussions wherein teachers create opportunities for students to share their thoughts, and then the teachers and students take up and work with those ideas. Indeed, the on-the-fly interactions that you have with students are a critical way for you to learn about the nature of student thinking. While these types of interactions might not be the first that come to mind when you think about tools in a formal process such as the Feedback Loop, we include them here for an important reason: The ways that you structure these interactions with your students for the purpose of revealing student thinking can also be thought of as tools. To create space for students to share their thinking, we've found that it's beneficial to talk openly and explicitly about how this type of conversation is different and to model and then identify for students how their participation in this type of classroom discourse is different.

Any of the written tools we listed above could be used as stepping stones for student thinking, so that students first respond to them and then turn to a partner or small group or participate in a whole-group discussion. At the same time, you could think about a whole-class discussion itself as a form of tool with a structure that you think about in advance.

Teacher-guided whole-class discussion: Michaels and colleagues (2013) have talked about formats for classroom talk in which the teacher guides a discussion around an initial problem or topic. The teacher allows students time to work out the issue at hand and take turns responding and, all the time, encourages them to listen to each other and build on each other's thinking. The *Accountable Talk Sourcebook* (*http://ifl.pitt.edu/index.php/download/index/ats*) includes concrete "talk moves" teachers can use to structure these discussions. These include challenging students by asking them, "What do *you* think?" which tosses the responsibility back to students (van Zee and Minstrell 1997), or recapping by asking students to summarize what they have learned. If you think of these types of talk moves as tools, you can see that—as part of the Feedback Loop—they can serve the purpose of surfacing student thinking in discussions around a particular goal. In addition, they use classroom talk norms that create space for students to participate equitably in the discussion, including using wait time after asking a question and after calling on a student as well as establishing turns for talk (so students are not speaking over each other; Michaels et al. 2013).

Small-group discussions: This discussion tool takes a particularly meaty or challenging question and poses it to a small group of students. Cohen (1994) argued that when asking students to work in groups, the question should be rich enough that it takes multiple students to dig into it. Once in a small group, multiple additional structures or tools can be used to guide student work; for example, you could ask students to come to a consensus on a particular response, forcing them to negotiate and trade off with different ideas. Conversely, you could simply ask students to have a discussion, and then to share out—either verbally or in some form of visual representation—the ideas represented in the group. If you're using a small-group discussion tool alongside a multiple-choice question, especially with clicker technology, you can group students together who have chosen different responses and ask them to take turns justifying their responses to each other.

CHAPTER 3

Think-pair-share: This is a simple discussion tool that teachers can use to break up a lesson and provide a chance to reveal student thinking around a particular goal. It's easy: Frame a question to students and give them a few minutes to *think* about their response. Then, *pair* students with the person next to them to discuss their ideas with each other. Finally, have students *share* out their ideas with the whole class. This strategy can work well to get students to preprocess their ideas before saying them to the whole class.

One teacher Erin worked with for several years preferred to lecture her students or have them do a guided activity with a worksheet, but she rarely provided space in her classroom for students to talk because she was afraid she would "lose control" of the classroom. With gentle support from her colleagues, she ultimately decided to try a think-pair-share about natural selection. Even though she had worked with her students in a traditional sense for an entire school year, she was delighted to see that, with this structure, students could talk to each other and share their ideas without her losing control of the class.

Concept maps: Although concept maps provide students opportunities to look at the relationships among different ideas, we've found them challenging to grade and assess. Teachers that Erin has worked with found that the discussions students had while making the maps were actually more interesting than the maps themselves; there is an example of this in Chapter 7.

Whiteboarding: This common strategy is a great way for students to share their initial and revised models for different phenomena because whiteboards are large enough for students to easily share their ideas with the class and with you.

Again, we suggest you complete Resource Activity 3.1 (if you haven't already); your ideas from the right-hand column can easily generate ideas for an assessment. The example in Resource Activity 3.1 takes up the idea of the transfer of matter and energy in an ecosystem through photosynthesis (MS-LS1-6), and an anchoring event for this can be found in the Private Universe video *Lessons From Thin Air*, which asks, "Where does all the mass in a tree come from?" This gets at the disciplinary core idea and crosscutting concept of MS-LS1-6 but not at the science practice. A follow-up prompt asking students to use evidence to support their explanation will get at this piece. A final tool might read, "Use evidence to support an explanation for where the mass in a tree comes from."

Designing, Selecting, and Adapting Tools

Supporting Goal–Tool Alignment With the Feedback Loop

Lane Carlson* is a first-year middle school Earth science teacher participating in a professional development program using the Feedback Loop as a tool to plan and reflect on instruction with other teachers. Mr. Carlson took the opportunity to reflect on an activity he had already designed and used with his students during an astronomy unit.

Mr. Carlson decided to complete a feedback loop for a classroom assessment he had recently made for his eighth-grade students during his astronomy unit. His original *goal* was to have students explore how the length of daylight and the angle of sunlight affect the seasons. To engage students, he had assigned them to a latitude and then asked them to use an online simulator to explore the times of sunrise and sunset and the Sun's angle at solar noon at different times of year. Mr. Carlson asked students to share their thinking about how the length of daylight and the angle of sunlight are related to temperature and then to connect that relationship with their prior knowledge of the Earth's tilted axis to explain the causes of the seasons.

He used some questions from a quiz a few days later as his *tool,* and the *data* from this included student responses to two multiple-choice questions and several true/false statements (see Figure 3.1, p. 48).

CHAPTER 3

FIGURE 3.1 Mr. Carlson's original tool

11. What are the two biggest reasons for warmer temperatures in summer?
a. <u>Closer distance to the Sun</u> and the <u>days are longer (more sunlight)</u>
b. <u>Closer distance to the Sun</u> and the <u>sunlight is more direct (higher angle)</u>
c. The <u>sunlight is more direct (higher angle)</u> and the <u>greenhouse effect is stronger</u>
d. The <u>sunlight is more direct (higher angle)</u> and the <u>days are longer (more sunlight)</u>

Directions: Indicate if the statement is true or false. If false, change the underlined words to make the statement true.

_____, _____ 12. On June 21, the Southern Hemisphere experiences the <u>longest</u> day of the year.

_____, _____ 13. The Earth has seasons due to its <u>distance from the Sun</u>.

_____, _____ 14. When the north end of the Earth's axis is tilted toward the Sun, it's <u>winter</u> in the Northern Hemisphere.

_____, _____ 15. In the Northern Hemisphere the spring equinox is followed by <u>longer</u> days.

_____, _____ 16. During winter in the Northern Hemisphere, Colorado receives <u>more direct</u> sunlight.

17. Label the following diagram with the four seasons *from the perspective of the Northern Hemisphere*.

A. _____
B. _____
D. _____

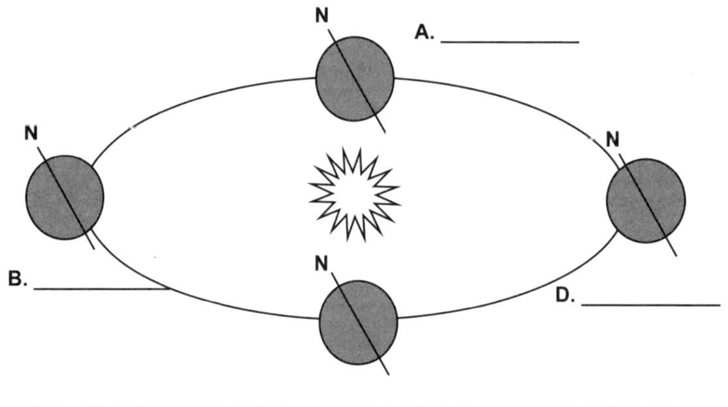

Designing, Selecting, and Adapting Tools

Mr. Carlson's reflection on his design and enactment of this activity is summarized in the feedback loop he completed during the professional development meeting (Figure 3.2).

FIGURE 3.2 Mr. Carlson's original feedback loop

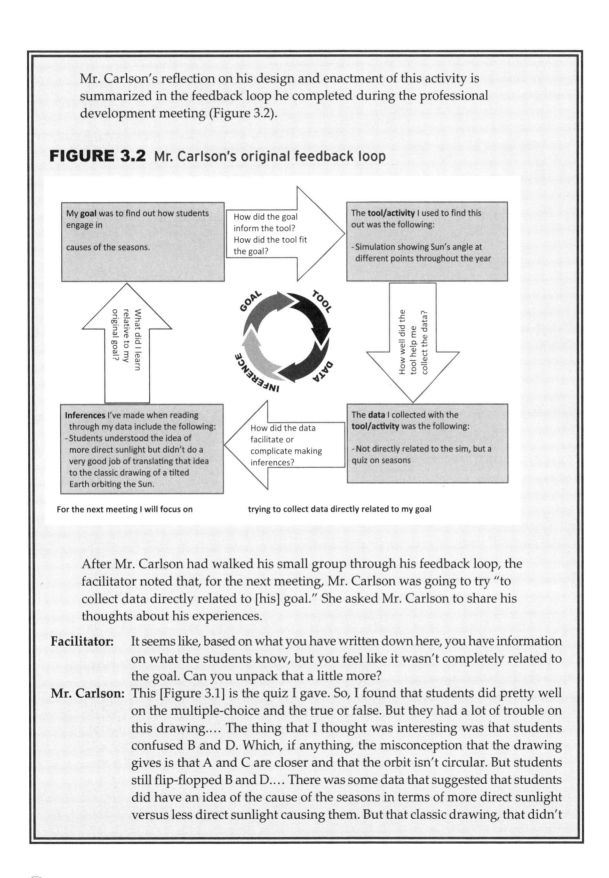

After Mr. Carlson had walked his small group through his feedback loop, the facilitator noted that, for the next meeting, Mr. Carlson was going to try "to collect data directly related to [his] goal." She asked Mr. Carlson to share his thoughts about his experiences.

Facilitator: It seems like, based on what you have written down here, you have information on what the students know, but you feel like it wasn't completely related to the goal. Can you unpack that a little more?

Mr. Carlson: This [Figure 3.1] is the quiz I gave. So, I found that students did pretty well on the multiple-choice and the true or false. But they had a lot of trouble on this drawing.... The thing that I thought was interesting was that students confused B and D. Which, if anything, the misconception that the drawing gives is that A and C are closer and that the orbit isn't circular. But students still flip-flopped B and D.... There was some data that suggested that students did have an idea of the cause of the seasons in terms of more direct sunlight versus less direct sunlight causing them. But that classic drawing, that didn't

CHAPTER 3

tell me what they got from the simulation because the simulation was all about the view from Earth.

Mr. Carlson also noted retrospectively that the quiz had focused on a disciplinary core idea of Earth and the Solar System: that the seasons are a result of the tilt of Earth's axis and the resultant intensity of sunlight on different areas of the Earth at different times of the year—without also focusing on modeling. *NGSS* MS-ESS1-1 states that students should "develop and use a model of the Earth-Sun-Moon system to describe the cyclic patterns of lunar phases, eclipses of the Sun and Moon, and seasons."

Mr. Carlson: I wonder if it would have been helpful to have the students use the physical models to create a drawing, to see what they would come up with. I wonder if that would be an interesting way to get their ideas first and then have them wrestle with the classic drawing.

Mr. Carlson decided to use the feedback loop for a subsequent goal and tool in the next unit, and then to collect data to guide inferences around that goal. His revised *goal* focused explicitly on the strengths and weaknesses of two different models in representing the solar system as well as explaining why models of the solar system are useful even when not completely accurate. Compared with his original goal, which was more about ideas and less about the practice of modeling, his new goal reflected the multicomponent nature of the *NGSS*.

The *tool* he used this time presented two different types of models to students: one that used the same scale for both sizes of and distances among the Sun and planets and another that used different scales. Mr. Carlson asked students to calculate the scaled diameters and distances using scale factors that he provided.

Designing, Selecting, and Adapting Tools

He then made scaled models out of butcher paper, and students paced out the scaled distances on the school's football field. Mr. Carlson instructed students to think about the strengths of each model as the class constructed them outside. Back inside, students were asked to describe the advantages of each model and explain which they thought was a better representation of the solar system. A sample of the *data* he collected is shown in Figure 3.3.

FIGURE 3.3 Mr. Carlson's second tool

> Now you will have to think about the pros and cons of the two models.
>
> - Model 1 - Same scale for planet diameters and distances between the sun and the planets.
> - Model 2 - Different scale for planet diameters and distances between the sun and the planets.
>
> 1. What does Model 1 clearly show us about the solar system?
>
>   *Model 1 clearly shows how extreemly far away the planets are from eachother, and how small they are in the big picture compared to eachother.*
>
> 2. What does Model 2 clearly show us about the solar system?
>
>   *Model 2 clearly shows how the planets sizes are compared to eachother.*
>
> 3. Which model best represents the solar system? Why did you chose that model as the best?
>
>   *Model 1 best represents the solar system. When the scale is the same for the size, the model is more accurate because you can visulize the extreem distance between the planets compared to the size of them. This is the most useful information because the model shows everything correct compared to eachother because of the scale factor that is used for both the distance and size.*

He was able to make more focused *inferences* on the basis of this new tool, including how well students grasped the idea that all the models were imperfect but that they could still be useful representations. Some students clung to the idea that the only "correct" model was the one that used the same scale for size and distance even though that model wasn't useful because it was so large

CHAPTER 3

(it wouldn't even fit on the school's football field). This second trip through the feedback loop is represented in Figure 3.4. After additional reflection, Mr. Carlson expressed the wish to revise his tool further:

I wish I had included a question to probe students' thinking around the usefulness, or utility, of each model. Some students may have answered that model 1 was a "better representation of the solar system," because they think models should be as accurate as possible, even at the expense of utility. But embedded in my goal was the desire to elicit students' ideas around the usefulness of models despite the model's weaknesses. In other words, model 1, while more true to reality in its scale, is not particularly useful as a model because it couldn't even fit on our football field, let alone on a textbook page or computer screen!

In reflecting on his first year of teaching, Mr. Carlson noted that one of his goals for his own teaching practice was to continue engaging in this process of bringing his tools into better alignment with his goals. He hoped that he could keep practices interlaced with ideas as he did so.

FIGURE 3.4 Mr. Carlson's second trip through the Feedback Loop

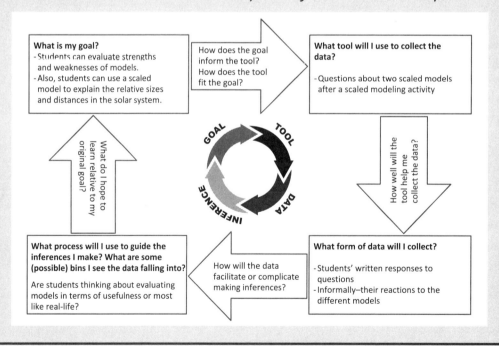

Tools for Assessing Science Practices

There is a clear consensus in science education circles that engaging students in science practices is essential (NGSS Lead States 2013). The National Research Council envisions that these three-dimensional goals will be assessed through multicomponent assessments (NRC 2014), although sometimes individual dimensions will, or possibly should, receive greater attention. For example, many traditional assessments foreground appraising what students know and are able to do in terms of disciplinary core ideas. Next, we focus on selected practices to discuss tools that could be used or developed to assess students' knowledge and abilities in each practice.

Asking Questions and Defining Problems

Students' skills with asking good questions and clearly defining problems might not be displayed through many traditional tests. Instead, this practice might more readily be seen when students observe demonstrations or videos, manipulate (or are introduced to) quantitative relationships among variables, or hear and read statements that other people present. Therefore, we need to ask ourselves how we can generate information about this practice during these instances. The strategy we use to generate this information is the tool.

One idea is to present a demonstration and have each student, or the class as a whole, record questions they have afterward about the outcome, its generalizability, and factors that might have affected what they saw. For example, when dropping an object at the same moment as you launch a similar object horizontally from the same height, students might note that the two items hit the ground at the same time; however, they might raise questions about whether the shapes of the objects matter or if the outcome would hold true regardless of the initial speed of the horizontal launch. Although it might be tempting to perform the demonstration and then *ask students questions yourself* (e.g., What did you see or hear? Why do you think that occurred?), it could be more valuable to have the students record questions, which could then be shared to see who is asking questions and what is being asked most often. This could lead students to plan and carry out investigations (another science practice).

On a related note, good questions can come up when students challenge or assess the design of an experiment. This can occur as a result of a demonstration, like the one above, or through lab exercises (or experiments the students read about or watch in videos). If one of your goals is to see whether your students can ask good questions, you can specifically have them record any new issues arising for them after doing the lab. To generate this information, you might need to modify your tools so that these questions are explicitly sought through students' lab reports or asked about in any handouts or files they complete along with a lab.

CHAPTER 3

In the examples above, the tools—or means of generating information about what students know and are able to do in regard to asking questions—include your expectations for posing questions as part of lab reports and questions you ask as you facilitate demonstrations.

Developing and Using Models

Models have enormous value in science and as science teachers, we want our students to gain greater understanding of how to develop and use models. To assess where students are in terms of this practice, you can ask them to create their own models when new phenomena are introduced. For example, when looking at predator-prey relationships, you can have students create a model of what's going on and ask them how the model would be affected as you introduce additional factors to the system. Their models, including initial ones and those changed as a result of including new information, can be useful for seeing students' understanding of how to use models. Before the lesson, it's important to think through what you're trying to learn and then develop tools—whether it's verbal instructions, explanations on a handout, or something else—that can best assist you with meeting that goal. You'll have to determine whether students will make their own models or work in pairs or small groups. Perhaps you'll want them to consider ways to represent their models using chart paper, colored markers, or other materials. Each of these decisions can affect the means of generating information about what students know.

As well as having students generate their own models, you can supply students with models and see how well they can use them in new situations. These could be models other students developed or ones you would like to share (e.g., ones that are often used in science classrooms or textbooks). The tools you use, including the homework, classwork, group work, or individual work students engage in, could ask students to use the model, and supply you with information about their understanding of what the model says and how to use it. Students could be asked to explore limitations or see what the model would suggest if applied to a novel problem. Your tool could push students to explore visual models, quantitative relationships among variables, three-dimensional representations, or verbal analogies that aim to model scientific ideas. As a way to tell you more about their understanding of what models are representing, you could even ask them to translate among different types of models (e.g., taking a verbal analogy and representing it visually).

In the examples above, the tools—or means of generating information about what students know and are able to do in terms of developing and using models—include your task and instructions for having them build representations of phenomena or apply models to novel situations.

Planning and Carrying Out Investigations

As science educators, we want our students to learn more about how to plan and move forward with investigations, both individually and collaboratively. It's important that they know how to carry out their explorations systematically, identify variables that might affect their outcome, and decide what information they wish to generate and how they will make sense of what they find. Time constraints associated with most traditional tests (e.g., limited to a one-day, 45–60 min., noncollaborative assessment) make it difficult to assess this practice through such a test.

Instead, to see what students know and are able to do in regard to planning and carrying out investigations, we could create multiday projects in which they work with others to investigate some phenomena. Perhaps you want them to explore why the number of fish in a nearby stream has dropped significantly over the last few years. They could be provided with materials and information about weather patterns and urban development; historical data about chemical levels in the stream; or other factors that might or might not have affected the fish population. Over several days, groups of students could perform tests on water samples and possibly interview people or research more factors that could play a role. The amount of time devoted to such an activity, as well as how much guidance or structure is supplied to your students, could vary depending on what's most appropriate for your setting. The students could ultimately create a poster and present their findings to others. Through this tool—the fish population activity itself—you could generate more information about their comfort in planning and carrying out investigations.

Similarly, you could grant students even greater flexibility in developing a science investigation. Science fairs or other avenues can be structured in ways that push students to plan and carry out investigations. You can have them provide periodic updates or create check-in points, to see what they know or are struggling with in this science practice and the content they are exploring. Additionally, any sharing of their work—whether through trifold brochures, slides, verbal presentations, or some other format—could have them show more about their plan and how they carried out their investigations. The work they do with the information they collect and analyze will also overlap with another practice, namely Analyzing and Interpreting Data. Depending on the work they do, or the expectations you build into this activity, students might provide additional information about other practices too.

In the examples above, the tools, or means of generating information about what students know and are able to do with carrying out investigations, include projects and assignments along with the expectations associated with them.

CHAPTER 3

Grounding the Tool in the Feedback Loop

Once you've come up with a draft of the tool, we suggest you spend some time anticipating how the activity will be enacted with students. Try the tool out yourself or share it with one or more of your colleagues to get their take. Think about how students might approach it, how they could interpret it, and the types of responses they might provide. If you are using any of the discussion tools described above, think about the question you will ask students, and the things they are likely to say in real-time. This practice, called Anticipating in the Five Practices model (Cartier et al. 2013), guides you to

> *actively envision how students might approach the instructional tasks or activities on which they will work. This involves much more than simply evaluating whether a task is at the right level of difficulty or of sufficient interest to students, and it goes beyond considering whether or not they are likely to get the "right answer." Anticipating students' responses involves developing considered expectations about how they might interpret a problem, the array of strategies—both correct and incorrect—that they might use to tackle it, and how those strategies and interpretations might relate to the concepts, representations, procedures and practices that the teacher would like his or her students to learn. (p. 28)*

This list of possible student responses will come in handy as you move ahead in the Feedback Loop to plan the forms of data you'd like to collect, as well as the inferences you will make from those data, and the feedback you will provide to students.

Involving Students

Framing assessment tasks in a variety of problem contexts will help students transfer their learning from one context to another (NRC 2014; Pellegrino 2001). Although we could guess at the problem contexts that might be intriguing to students, talking with the students themselves can generate rich ideas for the anchoring events (AST 2014; Yerrick 2000) in which to embed your tools. Often, these ideas can surface during informal conversations with students when they ask questions that apply what they're learning about to their everyday experiences. We've often been stumped by these questions initially, only to discover later that they make great problem frames for assessment tasks (e.g., a student's question of "Why can't we make Jell-O with fresh pineapple?" became a frame for studying enzyme function; a conversation about drought mitigation generates the musing, "I wish we could *make* it rain!" leading to a problem frame for cloud seeding, condensation nuclei, and the water cycle).

Furthermore, students can help determine the format of the tools you design. When Erin was a ninth-grade Earth science teacher, she wanted to get her students more involved in developing assessments, so she held student focus group sessions in which she created space for students to talk about how they liked to show her what they knew. They suggested a tool of a skit or using their bodies in some way. Erin was about to teach about laws of planetary motion, so she drew on this suggestion to design a tool in which students created a

Designing, Selecting, and Adapting Tools

physical model through a skit representing planetary bodies and motion. Students identified themselves as the various celestial bodies and sped up or slowed down as they got closer and farther away from a student representing the Sun. This allowed Erin to quickly assess students' knowledge of Kepler's laws.

Perhaps the most important factor for you to decide on when it comes to successful student participation in a feedback loop is how much to involve them in the process of setting norms and expectations for their participation. Do students feel safe to be open and vulnerable about their learning and understanding? Is it OK for them to say, "I'm not sure," or "I don't understand?" How do other students respond when someone says they're struggling? Regardless of the tool you choose, it needs to be used in a classroom climate where students feel it's safe to share their thinking.

Having a positive space for formative assessment obviously does not happen by accident. As the teacher, you have the opportunity to model and develop the norms in your classroom explicitly. Rather than assuming students already know how to share and discuss their ideas in a classroom setting, you can specifically teach and allow students to practice the skills necessary to take charge of their learning.

In the following vignette, a teacher, Alice Schafer*, holds a whole-class discussion with her sixth-grade students about how she is listening to them in class. She elicits her students' observations about her, as the teacher, after they just engaged in a discussion about sinking and floating. To have them analyze the conversation, she says, "I want to talk about how I've been listening to your ideas. Can you tell me what you've seen about how I listen to any idea you've said?" Then, the students respond with their perceptions of the "teacher moves" she makes to elicit conversation and discussion. By taking the time to point out some of the strategies she uses as she listens to the students, she is modeling and showing students some of the strategies *they* can use to listen to their peers.

CHAPTER 3

Setting Norms for Formative Assessment Conversations

Ms. Schafer*, a middle school science teacher, was dedicated to creating a classroom climate conducive to formative assessment. She did not expect all her students to know automatically how she expected them to participate in a whole-class discussion, so she dedicated class time to helping students identify how she listened to their ideas and how she wanted them to talk with each other. She described her own teaching as follows:

I try really hard to model that all of the time. You know, this is our problem. What's our problem? Do we know anything about it? All right, what's your hypothesis? What do you think is going to happen? And then always referring back to those steps with our procedure. Where are we going? Why are we doing this? ... What's this about? Why do I need to do this? I like to let them in on what I'm doing, because if I don't intentionally inform them about the process and what I'm doing, why would they ever ... why would I assume that they would know why I'm doing what I'm doing?

One of Ms. Schafer's classroom practices that she used to help students understand how to participate in formative assessment was to have a whole-class conversation about how to share ideas. The first time she held an assessment conversation with her students, she gathered them together and invited them to talk about how she listened, holding herself as a model of how she would like students to treat each other. In this way, Ms. Schafer created an opportunity for students to identify her actions that supported the students' participation in formative assessment (all student names are pseudonyms).

Ms. Schafer: Let's have a conversation. I'd like you to close your books, and I'll put away my books. I want to talk about how I've been listening to your ideas. Can you tell me what you've seen about how I listen to any idea you've said? How do I listen to your ideas? Susanna?
Susanna: After we say it, you repeat our ideas so everyone can hear it.
Ms. Schafer: After you say it, sometimes I repeat your idea so everybody can hear.

Here, Susanna immediately identified Ms. Schafer's regular use of a strategy called *revoicing*, for which the teacher repeats what the student said back to him or her to check to see if the student has been understood correctly (O'Connor and Michaels 1993).

Brittney: You listen intently.
Ms. Schafer: Oh, I listen intently.

Designing, Selecting, and Adapting Tools

Kenzie: And you make sure everyone else is listening.
Ms. Schafer: So I listen intently. I'm focused on you.
Kenzie: Yeah, and [respect what we're saying].
Ms. Schafer: So I make sure everybody's listening to you, and I respect what you're saying.
Student: You look at us.
Ms. Schafer: I look right at you when you're talking. OK. Ginny.
Ginny: You don't tell people that their answers are wrong.
Ms. Schafer: I don't tell people their answers are wrong. Yeah. I don't, do I? OK, Henry?
Henry: You're not strict. You're really nice.
Ms. Schafer: Sometimes I'm not strict, and I'm really nice. Yeah.

While this might seem obvious, having a demeanor in the classroom where students don't feel like their ideas are going to be immediately judged and where students feel that the teacher is kind and listens carefully is key to creating an atmosphere in which students can share their ideas, even if they are feeling uncertain about them or if they are worried they might be wrong.

Sammy: You call on everybody. When you start calling on one person, you let everybody get a chance to talk.
Ms. Schafer: So everybody gets a chance to talk. That is important. I just don't call on one person.

Sammy has noticed that Ms. Schafer makes sure that all of the students are able to talk. In her classroom, she does this through a routine of hand-raising and clear turn-taking so that students are not talking over each other, and so she is monitoring and soliciting contributions from all her students (Michaels et al. 2013).

Jacque: I don't know if someone already said this, but you have [us write things] on the board, like the ideas.
Ms. Schafer: Oh, I let other students write things on the board. Do you like that?
Jacque: Yeah.

Jacque noted that Ms. Schafer uses her whiteboard democratically, providing students opportunities to write their own ideas and to make those ideas public so that the whole class can view them and work with them. After this sustained conversation, Ms. Schafer turns to the work at hand, saying, "I would like you to share with me your ideas, respectfully. Treating each other like I've treated you."

In this conversation, which took less than 10 minutes, Ms. Schafer created space in her classroom—and engaged students in surfacing a set of norms—that set up a climate in which students felt safe to share their ideas (they did not have to worry about being evaluated) and in which their ideas would be made public and shared with the rest of the class. In effect, by asking students to name the talk moves that she modeled daily, Ms. Schafer was able to involve them in identifying norms that were conducive to holding an assessment conversation.

CHAPTER 3

Summary

The information above provides ideas for tools teachers could use to explore selected science practices in their classrooms. The tools mentioned include ways to facilitate demonstrations and related discussions, expectations you can develop for lab reports, instructions for working with models, and projects that push students to plan and pursue scientific investigations. They do not represent a complete list of all the potential tools that could be used to explore each practice. Instead, they highlight various tools available to us as teachers and how those tools could be useful for exploring different practices and goals. While we only discussed three of the *NGSS* practices, similar work could be done with the remaining science practices. We encourage you to use resource activities at the end of this chapter to brainstorm ideas and sources for tools, and then to evaluate the quality of those tools.

Designing, Selecting, and Adapting Tools

How Do We Know If Climate Change Is Happening?

An Example of a Multicomponent Tool

In the summer of 2009, a group of middle school science teachers came together to develop activities and assessments to support student learning about climate change in Colorado. Although this workshop was conducted pre-*NGSS*, the focus of teachers' design work was to develop tools that would interweave ideas about climate change with the science practices of analyzing and interpreting data, as described in HS-ESS3-5. Their goal was to determine how students drew on various sources of data to develop explanations and to use those data to determine if climate change is occurring in Colorado.

Four teachers (Laura Bicknell, Courtney Kellie, Jessica Campbell, and Jon Gerber) collaborated with climate scientists from the University of Colorado Boulder who had collected and represented multiple forms of data illustrating changes in climate in Colorado over the past 100 years. The teachers hoped to engage students in the science practice of analyzing data with these actual representations. Students need opportunities to analyze large data sets and identify correlations. These types of data sets are increasingly available on the internet (NRC 2012); for example, the Bguile project assembled by Brian Reiser and colleagues at Northwestern (*http://bguile.northwestern.edu/introduction.html*) offers opportunities for teachers and students to create representations of Peter and Rosemary Grant's original data from Darwin's finches in the Galápagos Islands.

Collaborating with university education researchers, the teachers identified a set of representations of climate data that they wanted to provide directly to their students for interpretation. These data included historical temperatures at multiple locations in Colorado, snowpack, forest fire frequency, and images of local glaciers over time; the original data sheets are available at *http://learnmoreaboutclimate.colorado.edu/lessons/view/id/6*.

To scaffold this assignment, the teachers developed a set of tools. The first was to guide students' interpretation of each piece of data; for students in lower grades, groups of three to four would each be provided with one piece of data. Students were led through their interpretation of that piece of data with the questions in the tool below (Figure 3.5, p. 62).

CHAPTER 3

FIGURE 3.5 Tool guiding students' data interpretation

Procedure

1. Read the data packet. Does your evidence support climate change? Why or why not? Record ideas in in the table below. This should take about five minutes.

2. Join your group. Share your thoughts. Be sure to listen to the ideas of others. Record your group's observations in Figure 3.6.

3. Did you all agree? If not, keep this discussion going. Try to resolve your differences and reach a consensus. Record final results in results section.

4. Be prepared to present your findings to the class. State your consensus. Each member should be able to give the class a reason why you agreed.

Observations

Does the evidence support climate change? (yes or no)	Why or why not?

The second tool the teachers developed then guided students in recording the observations of their group members (Figure 3.6).

FIGURE 3.6 Tool summarizing sources of data and reasoning from evidence about climate change

Data source	Yes or no	Evidence

Finally, students reflected on all the different pieces of data to determine whether or not they thought that the evidence supported the argument that climate

CHAPTER 3

> change was occurring in Colorado. The group had to explain at least two reasons why they reached their consensus.
>
> Taken together, these tools combined multiple components: original qualitative and quantitative data about weather and conditions in Colorado, a guide to interpreting that data, a scaffold to support students' reasoning about those data, and constructed-response questions asking them to come up with a scientific explanation on the basis of those data to support a claim as to whether climate change was happening in Colorado. The different components of the activity supported different parts of the goal, providing multiple sources of data to help teachers make inferences about student learning.

Designing, Selecting, and Adapting Tools

RESOURCE ACTIVITY 3.1
Brainstorming Features for Multicomponent Formative Assessment Tools

Unpack the NGSS performance expectation you're working with into the science practice, disciplinary core idea, and crosscutting concept. Then brainstorm ways that a multicomponent tool would need to draw out student thinking and practices for each element.

SAMPLE

Standard or performance expectation
MS-LS1-6: *Construct a scientific explanation based on evidence for the role of photosynthesis in the cycling of matter and flow of energy into and out of organisms.*

NGSS DIMENSION	WHERE IS THIS ELEMENT PRESENT?	HOW MIGHT I ASSESS THIS IN A TOOL?
Science practice	Construct a scientific explanation based on evidence	Ask students to explain a phenomenon (e.g., Where does the mass in a tree come from?); provide scaffolds to remind students of elements of an explanation.
Disciplinary core idea	Photosynthesis	Have students identify the chemical reaction by which plants produce energy by "fixing" carbon from the atmosphere.
Crosscutting concept	Energy and Matter	Look for students relating the Sun's energy to the matter created through photosynthesis.

YOUR TURN

Standard or performance expectation

NGSS DIMENSION	WHERE IS THIS ELEMENT PRESENT?	HOW MIGHT I ASSESS THIS IN A TOOL?
Science practice		
Disciplinary core idea		
Crosscutting concept		

CHAPTER 3

RESOURCE ACTIVITY 3.2
Evaluating the Quality of Formative Assessment Tools

CRITERION	QUESTION	HOW DOES THE TOOL FIT THIS CRITERION?	WHAT NEEDS TO BE MODIFIED?
Alignment with goals	What do you want students to know and be able to do?		
Focused on Big Ideas	What big idea or anchoring phenomenon does the tool address?		
Surfaces student thinking	What kinds of questions does the tool use to surface student thinking?		
	What types of responses do you expect the tool to surface?		
	Is the tool eliciting extraneous information that is unrelated to the goals you have for it?		
Easily interpretable information	What type of information about student ideas does this tool provide?		
	How much time or energy do you anticipate it would take to interpret the data that this tool will generate?		

Designing, Selecting, and Adapting Tools

RESOURCE ACTIVITY 3.3
Anticipation Guide

Once you have drafted a complete tool, take a few moments to anticipate the different ideas you expect to surface with each part and envision how you might respond if students share those ideas.

POSSIBLE RESPONSES STUDENTS MAY WRITE OR SHARE	WHAT I WILL DO

CHAPTER 3

References

Ambitious Science Teaching (AST). 2014. Planning for engagement with important science ideas. Seattle, WA: University of Washington Department of Education. *http://ambitiousscienceteaching.org/wp-content/uploads/2014/08/Primer-Plannning-for-Engagement.pdf*

Black, P., and D. Wiliam. 1998. Inside the black box: Raising standards through classroom assessment. *Phi Delta Kappan* 80 (2): 139–148.

Briggs, D. C., A. C. Alonzo, C. Schwab, and M. Wilson. 2006. Diagnostic assessment with multiple-choice items. *Educational Assessment* 11 (1): 33–63.

Cartier, J. L., M. S. Smith, M. K. Stein, and D. K. Ross. 2013. *5 Practices for orchestrating task-based discussions in science*. Arlington, VA: NSTA Press.

Cohen, E. G. 1994. *Designing groupwork*. 2nd ed. New York: Teachers College Press.

Duschl, R. A., and D. H. Gitomer. 1997. Strategies and challenges to changing the focus of assessment and instruction in science classrooms. *Educational Assessment* 4 (1): 37–73.

Furtak, E. M. 2009. *Formative assessment for secondary science teachers*. Thousand Oaks, CA: Corwin Press.

Grossman, P. L., P. Smagorinsky, and S. Valencia. 1999. Tools for teaching English: A theoretical framework for research on learning to teach. *American Journal of Education* 108 (1): 1–29.

Kang, H., J. Thompson, and M. Windschitl. 2014. Creating opportunities for students to show what they know: The role of scaffolding in assessment tasks. *Science Education* 98 (4): 674–704.

Keeley, P., F. Eberle, and L. Farrin. 2005. *Uncovering student ideas in science, volume 1: 25 formative assessment probes*. Arlington, VA: NSTA Press.

Li, M., M. A. Ruiz-Primo, and R. J. Shavelson. 2006. Toward a science achievement framework: The case of TIMSS-R study. In *Contexts of learning mathematics and science: Lessons learned from TIMSS*, ed. T. Plomp and S. Howie. 291–312. New York: Routledge.

Michaels, S., M. C. O'Connor, M. W. Hall, and L. Resnick. 2013. *Accountable talk sourcebook: For classroom conversation that works*. Pittsburgh, PA: University of Pittsburgh Institute for Learning.

National Research Council (NRC). 2012. *A framework for K–12 science education: Practices, crosscutting concepts, and core ideas*. Washington, DC: National Academies Press.

National Research Council (NRC). 2014. *Developing assessments for the Next Generation Science Standards*. Washington DC: National Academies Press.

NGSS Lead States. 2013. *Next Generation Science Standards: For states, by states*. Washington, DC: National Academies Press. *www.nextgenscience.org/next-generation-science-standards*.

O'Connor, M. C., and Michaels, S. 1993. Aligning academic task and participation status through revoicing: Analysis of a classroom discourse strategy. *Anthropology and Education Quarterly* 24 (4): 318–335.

Pellegrino, J. W., N. Chudowsky, and R. Glaser. 2001. *Knowing what students know: The science and design of educational assessment*. Washington, DC: National Academies Press.

van Zee, E., and J. Minstrell. 1997. Using questioning to guide student thinking. *The Journal of the Learning Sciences* 6 (2): 227–269.

White, R., and R. Gunstone. 1992. *Probing understanding*. New York: Falmer.

Yerrick, R. K. 2000. Lower-track science students' argumentation and open inquiry instruction. *Journal of Research in Science Teaching* 37 (8): 807–838.

CHAPTER 4

Collecting Data

I have data! Now, how should I use it?

—First-year high school chemistry teacher

Say the word *data* to a science teacher and certain images may come to mind, such as spreadsheets, tables, graphs, or lists of numbers. These forms of data are part of the everyday experience of practicing scientists. Depending on the type of science we're talking about, data might also include field notes of animals' behavior, drawings, maps, or samples such as tree and ice cores, rocks, or blood. Our backgrounds in science lead us to call these forms of data most immediately to our attention.

At the same time, the word data is floating around educational reform circles. Everywhere you turn, it seems as if some form of data is being collected and then used as a foundation for a new catchphrase policy; for example, "data-driven instruction" or "data-driven decision making." A recent New York Times article (Rich 2015) summarized this trend:

> *Custodians monitor dirt under bathroom sinks, while the high school cafeteria supervisor tracks parent and student surveys of lunchroom food preferences. Administrators record monthly tallies of student disciplinary actions, and teachers post scatter plot diagrams of quiz scores on classroom walls. Even kindergartners use brightly colored dots on charts to show how many letters or short words they can recognize.*
>
> *Data has become a dirty word in some education circles, seen as a proxy for an obsessive focus on tracking standardized test scores. But some school districts, taking a cue from the business world, are fully embracing metrics, recording and analyzing every scrap of information related to school operations. Their goal is*

CHAPTER 4

to help improve everything from school bus routes and classroom cleanliness to reading comprehension and knowledge of algebraic equations.

As educators, we think data are critically important, too, which is why it's one of the four elements of the Feedback Loop. However, we see data as serving the purpose of improving the quality of instruction through an aligned process. We use data to refer specifically to the multiple forms of evidence generated by our tool that will then guide inferences about whether or not students have met the goal. This chapter will talk about multiple types of data in the Feedback Loop and ways of reducing the amount of data you collect.

The Role of Data in the Feedback Loop

As the *New York Times* article described, the term *data* often conjures the results of standardized tests that are delivered externally and are often difficult to translate into prescriptions for action in the classroom. This view doesn't just see this particular type of quantitative data as superior, it sees all data collected for accountability purposes as superior to any collected independently by teachers, regardless of type. As we turn to discuss the role of data in the Feedback Loop, we will broaden the common idea of data as being external to the teacher and primarily consisting of standardized test scores. In contrast, data in the Feedback Loop are constituted by the multiple sources of information about student thinking that are generated by your tool (see Chapter 3). The tools you select and design to align with your goals should in turn generate data that are useful to you in determining what students understand and are able to do. Just as the form of the tools varies, so can the data those tools generate.

Before we get too far, we'll add a word of caution: Some teachers who have worked with us and the Feedback Loop have been confused by our distinction between the tool and the data that the tool generates. We'll work to be very explicit throughout this chapter as to what we mean by these two elements.

Qualitative and Quantitative Data

Classroom data are most commonly grouped into two categories: qualitative and quantitative. As teachers and as scientists, we know the value of both for different purposes and how both can work independently and together. Quantitative data include anything that can be counted. This could be standardized test scores, class quiz results, or even something as simple as the number of students who answered a warm-up question correctly. Quantitative data can be manipulated statistically and represented in charts and graphs to more easily show trends and patterns. The data could be about the whole class or about a specific student. You might use quantitative data when you are trying to quickly assess the understanding of the class overall or of a particular student, to determine where you might want to explore more deeply.

Collecting Data

We tend to be a bit dubious of the traditional "85%" reports teachers can get about how students did on tests and quizzes, because these numbers tend to aggregate a lot of information about what students know and are able to do. If a student has an 85% level of understanding, what does that actually mean? The standards-based language of proficient, partially proficient, and so on gets more at information about what standards students have met and what they still have to learn; however, it is only useful if the information is reported in relation to standards. If you are able to generate reports of this nature, such that your test or quiz is neatly aligned with the goals you set, then the numbers your test generates will be more useful. If not, it might be beneficial to consider alternative forms of data.

In contrast, qualitative data seem to get much less airtime in education reform. They may not have the same perceived level of validity or street credibility as quantitative data, but they have a key place in the Feedback Loop. The problem with the assumption that qualitative data are less useful is that much of the data generated in a classroom daily—including students' written responses, their expressions, and their questions and contributions to whole-class conversations—are qualitative. As a result, if we are favoring quantitative data, we have to find a way, and time, to convert qualitative data to quantitative data by grading and assigning points to everything. Information is inevitably lost in this process, and the delay between when students generate data and when you're able to look at it can get in the way of your instruction being responsive to student thinking.

Ironically, for some science teachers, the idea of using qualitative data in the classroom is baffling. As scientists, they are much more comfortable with facts, numbers, and other data that may be deemed more objective. However, by looking at different types of qualitative data, teachers can get a better sense of how students are thinking and not just see whether or not they "get it." This can be extremely important, particularly in teaching science, because teachers must address preconceptions, partial conceptions, and misconceptions if students are to develop strong understandings in their science learning. By uncovering how a student thinks about a particular concept, you can diagnose and identify what pieces are missing and specifically target your instruction to meet your learning objectives.

This is why we really like using qualitative data in the Feedback Loop. It is closer to what you do every day, and without creating the need to score things and assign points, we open up a whole world of possibilities. In fact, research has shown that it's better *not* to score student responses if you want to use them for a formative purpose; it doesn't just slow down the time to give feedback, but students have been shown to disregard qualitative feedback in favor of looking at their grades anyway (Butler and Nisan 1986).

Let's work through a couple of examples of tools and the data they capture and generate. An ecologist might make field notes in a waterproof notebook about the behavior of pikas (a small mammal living at high altitudes), which are a qualitative form of data. Her observations of the pikas are assisted by the use of binoculars, so she can observe the mammals from afar.

CHAPTER 4

At the same time, this ecologist might quantify the number and location of pika burrows on rocky mountainsides. She has used several different tools: a pen, a waterproof notebook, a pair of binoculars, perhaps a GPS receiver to note the locations of the pika burrows, and a physical map or computer program to record those locations. These tools in turn help her generate and capture several forms of data—field notes and burrow numbers and locations—which together provide rich information about the behaviors and population density of the pikas.

Similarly, in the classroom, the tools we use can similarly generate quantitative and qualitative data about student learning. An environmental science teacher might engage her students with a physical model of erosion using stream tables. The tools she uses include the activity she has developed to guide students to focus on the relationships among variables that influence stream velocity, slope, and characteristics of the water corridor such as meandering. The tools also include the stream table itself, sand, and the water students channel through the sand. As students engage with them, these tools combine to generate multiple sources of data in qualitative or quantitative form that the teacher can collect. As students model different geological features, she notes students' expressions of confusion or frustration while they're working. To be sure that students are engaging in the main goals of the activity, she regularly drops into groups to ask about the variables they are changing and the effects of shifting those variables, such as the velocity of the water or the angle of the stream table, on the features of the riverbed they are creating. At the end of the activity, she collects students' descriptions of the relationships modeled with the stream table and scores them with a rubric. The teacher here has collected multiple sources of data, including the qualitative expressions on student faces and responses to her questions and quantitative responses to the activity via her rubric.

In each of these instances (the ecologist and the environmental science teacher), we can see illustrated the distinction between tools and data. The binoculars the ecologist uses are not the same as the observations she makes when using them, just as the activity that students use to guide their modeling of the stream system is not the same as students' responses to that activity. Similarly, the tools can be used in different ways to generate different types of data: the ecologist may look through the binoculars to gather qualitative descriptions of pika behavior or she may use those same binoculars to identify burrows, which she tallies and later enters into a spreadsheet. The environmental science teacher may simply read through student responses to the activity, noting the nature of student ideas and picking up on themes to visit in subsequent lessons; she may also score those responses according to their accuracy.

In the following sections, we will give several examples of the different types of quantitative and qualitative data that formative assessment tools can generate. Ultimately, the type of data that any tool generates depends on the teacher, and how she or he determines to enact a given tool with students. Just as a multiple-choice question is easily scored (quantitative), it can also generate rich class discussions (qualitative).

Collecting Data

Initial and Revised Models in the Feedback Loop

Kate Henson, Miss Porter's School, Farmington, Connecticut

I have been teaching biology since 2001 and have worked in public, charter, and independent schools. Currently, I am teaching at Miss Porter's School, an all-girl, independent day and boarding school in Connecticut. Our student body of 320 students includes US and international students and is socioeconomically and ethnically diverse.

The Feedback Loop really resonated with me because it named the process I have been using for years, although I had never explicitly thought about the steps before. I used the Feedback Loop to formalize my process of designing, enacting, and reflecting on a formative assessment in one section of biology. This year-long course is the second in the required physics–biology–chemistry sequence. This particular class comprised 11 students, including 1 sophomore, 8 juniors, and 2 seniors. Students had already completed a unit on biochemistry before studying cellular biology.

Within the context of cellular transport, my *goal* was to see students model and explain how substances move across the cell membrane. I knew the students had a good working knowledge of the structure and function of the cell and cell membrane, but I hoped that the use of whiteboard models would allow me to understand how they were applying what they already knew about molecules and cells to a new situation.

I developed a simple *tool* to elicit student ideas prior to engaging them in classroom experiences related to transport: a quick set of instructions that I jotted onto my whiteboard. I asked students to create a pictorial model of what they thought would happen to plant cells in different scenarios (Figure 4.1, p. 74).

CHAPTER 4

FIGURE 4.1 The tool prompting students to draw initial models showing the movement of water in and out of a cell

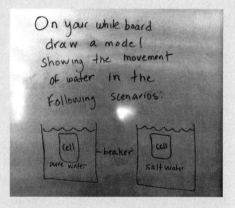

In the first scenario, the plant cells would be put in a saltwater solution, and in the second, they would be put in a deionized water solution. I divided my students into groups of two to three and gave each group a small whiteboard (2′ × 3′) and several dry-erase markers.

The student-drawn models on the whiteboards provided me with the *data* I needed to understand their thinking and determine which classroom experiences should follow. Figure 4.2 shows two sample initial models. In the model in Figure 4.2a, one pair of students predicted that in both scenarios water would flow in and out of the cell, but in the saltwater solution the rate would be slower because salt would block the channels and prohibit the water's movement. In the pure water solution, their model indicated there would be more water flowing in than out but did not indicate why. In the model in Figure 4.2b, another pair of students predicted that in the pure water solution, equal amounts of water would flow in both directions, resulting in no net change, whereas in the saltwater solution more water would flow out than in. They reasoned that the flow of water into the cell would be slowed down by the smaller Na and Cl ions, which would have an easier time flowing into the cell.

Collecting Data

FIGURE 4.2 Sample student initial models: (a) Water flows in and out of the cell, but the rate is slower in salt water. (b) In pure water, equal amounts of water flow in both directions and in salt water, more water flows in than out.

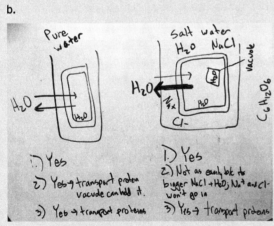

The data from these whiteboard models led me to make several *inferences*. Students could predict that water would move both in and out of the cell. They understood that the cell membrane is permeable to water, but they weren't sure how much water would be flowing in which direction and why.

This led me to set the new *goal* of providing students with an experience that would allow them to revise their original models into an accurate working model. My new *tool* was an activity in which students made wet mount slides of Elodea leaves and looked at them under the microscope. Once the students made their slides and located the cells under the microscope, they treated the cells with a saltwater solution and recorded their observations. They then repeated this procedure with a deionized water solution before discussing results with their partners and revising their models.

These revised models, along with students' descriptions of the models they shared in a whole-class discussion, formed a new source of *data*. Each pair took a turn projecting their modified original models on the whiteboard at the front of the room. Figure 4.3 (p.76) shows revisions students made to their models in lighter-colored ink.

CHAPTER 4

FIGURE 4.3 Revised student model showing how water moves into and out of cells in pure or saltwater solutions

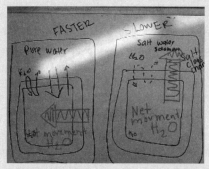

The new model showed the "water arrow" pointing in when the cell is placed in pure water and out when placed in salt water. Students explained that the water "wanted to go where there was less water." This data led me to the make the *inference* that although they may not have been using scientific terms to describe their observations, they were able to construct an accurate model.

No method of learning is linear. We never get to the point where we know everything and stop; that's what's so great about the Feedback Loop. We get new information and go around again. The Feedback Loop gave me a framework to work in, both for my original goal of uncovering student ideas and models for membrane transport and for the new goal of engaging students in an investigation to help them revise their models.

For me, the Feedback Loop was a powerful tool. In addition to helping me revise my goals, it helped me easily construct meaning from the information I was getting from my students. It forced me to listen carefully. As an experienced teacher, I have accumulated a lot of tools, and it's easy to choose an experience for students for the wrong reasons. For example, I might really enjoy a particular lab and think it's fun for my students, but that doesn't make it meaningful or the right experience for where my students are in terms of their understandings. The questions the Feedback Loop made me ask myself are, "What is my goal?" and "What tool will help me reach my goal?" Then, once I had the data elicited from my students, I asked, "Did it work? Did my tool help me reach my goal?" And then, considering my inferences, I asked myself, "Where do we go next? What is my next goal?"

Collecting Data

Formal and Informal Data

Data span the continuum from formal to informal (Cowie and Bell 1999; Ruiz-Primo and Furtak 2006, 2007). Formal types of data are usually the result of tools planned in advance; such tools are often handed out to students or shown to them on a screen. Formal data fit into a lesson plan and constitute information you might share with your colleagues (Ainsworth and Viegut 2006). They capture the nature of student thinking at a particular point in time such that you might look back at them later to make inferences to guide your instruction.In contrast, informal data include students' responses to questions asked on the fly, their expressions, and their participation in class. This is what experienced teachers might simply call "good teaching." Informal data are often generated through discussion tools, such as when you ask questions, listen to small-group discussions, pay attention to students' participation or their tone of voice, or do anything else that helps you keep track of how a lesson is going.

To illustrate the difference between formal and informal data, let's return to the environmental science teacher from earlier who collected formal and informal data in addition to the qualitative and quantitative data already discussed. She collected informal, qualitative data about student expressions and participation during the lab. The questions she asked students about the variables were planned in advance, but she also asked improvised questions, making it a semiformal or planned source of data. Finally, she scored students' qualitative, formal responses to the activity that guided their modeling with the stream table. We summarize all these different forms of data in Figure 4.4.

FIGURE 4.4 Summary representation of types of classroom data

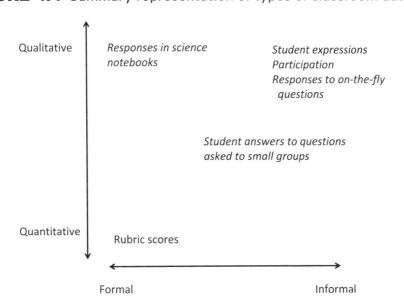

CHAPTER 4

Although not all forms of data fall cleanly into this formal/informal, qualitative/quantitative representation, we do find it to be a useful way to think about the types of data we typically collect in our classrooms. If the majority of data you collect in feedback loops are located at the quantitative-formal intersection—aggregating student clicker responses in graphs, for example—it would be worth thinking about how more informal-qualitative forms of data could be generated to complement these, perhaps by listening to discussions by students of their responses in small groups.

Data Reduction

One of the most common things we hear from teachers when we work with them on the Feedback Loop is how overwhelmed they are with data. "What do I do with all this data? I'm swimming in it, and you want me to collect more?" Ironically, now that we've broadened the idea of data, we'd like to talk about the process of data *reduction*. That is, we want to guide you through the process of using the Feedback Loop to help you reduce the overall amount of data you are collecting and looking at daily.

The phrase "data reduction" comes from qualitative social science research, which puts reducing the amount of data you're working with as the first step in moving toward drawing and verifying conclusions. This process has also been called condensation because it refers to "the process of selecting, focusing, simplifying, abstracting, and transforming the data" (Miles and Huberman 1994, p. 12). This needs to be done to make data more manageable, as well as more interpretable relative to your original goals.

If you're like we were before we began using the Feedback Loop, your daily practice involves collecting tons of data. Erin has memories of lugging multiple canvas bags home every night, filled with numerous six-inch tall stacks of student work from her five classes every day. She hauled bags often enough that on teacher appreciation day, when she was getting a free shoulder massage, the therapist immediately identified which shoulder she usually carried the bags on. Zora remembers stacks and stacks of collected homework assignments that taunted her every Sunday night as she frantically tried to catch up and enter grades for weekly progress reports. However, how much of this information was actually useful in adapting our instruction to help students meet our learning goals?

One of the major purposes of the Feedback Loop is to help you be more deliberate about the data you collect. When we talk to teachers who feel overwhelmed by the data they are collecting, we find they are stuck in that lower-right hand corner of the loop. Usually, when we press these teachers to reflect on the other vertexes of the loop, they are not clear on what goal they have in mind, and the tools they use are so long that they generate much more data than necessary. Then, looking forward in the loop, these same teachers struggle to make inferences because they have so much to look at and are not sure where to begin. Figure 4.5 represents the relationship of data to other elements of the Feedback Loop.

Collecting Data

FIGURE 4.5 Data in the Feedback Loop; lighter-colored arrows highlight links between data and other elements of the loop

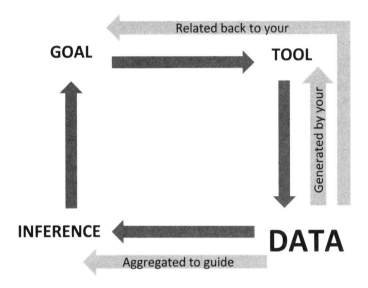

We suggest starting data reduction by going through the following process: Think back to the goal that you had and the tool you used and streamline. If you asked students five questions on a written tool, which of them best aligns with the goal? If you can identify that one question, then you've just reduced your data by 80%.

Sometimes we get a little confused ourselves when talking with teachers about the difference between data and tools. This can be very straightforward for formal sources of data, such as if you design an activity (*tool*) and give it to students to record their written answers (*data*). However, it can get a bit confusing when we're talking about informal forms of data, such as class discussions.

Howard notes that how he would find something is often through information. For example, he could say *how* he would find out what students know is through having them engage in discussions (which he might facilitate), yet the *information* he would look at is the discussions themselves. So, discussions are both how he would find out what students know and the information about what students know. In other words, in-class discussions can contain both the tool and the data.

This might be more clearly seen when looking at something such as a video recording: A teacher could say that how he or she would find out what students know is through making a video of them, while also saying that the information he or she would look at is

CHAPTER 4

> the video itself. So making a video is *how* the teacher would find out what students know, and it contains the information about *what* students know. In other words, the video can also be both the tool and the data.

Involving Students

Student involvement in the Feedback Loop is perhaps most straightforward when we are talking about data: They are the ones creating the data about what they know and are able to do. If we take a limited view of data, this information is something students generate in one step and you look at in another step. However, we've found that actively engaging students in generating data and listening and responding to the ideas (data) of other students can enrich the feedback loop and generate data in real-time.

In this chapter, we presented the approach of the assessment conversation as a simple and fast method of actively having students share their ideas in real-time (Duschl and Gitomer 1997). If you engage students in conversations like these, over time they can even learn to take the role that teachers traditionally occupy, holding their peers accountable to their ideas and challenging and pushing each other in their thinking (e.g., Engle and Conant 2002).

Another approach to engaging students in selecting data for you is by having them assemble portfolios in which they choose what they think counts as high-quality data of their learning. If you are using a running-record source of student data, such as science notebooks (e.g., Chesboro 2006), students could even identify which pages they felt were the best representation of their work and direct you to look at those pieces of evidence for their learning.

Summary

In this chapter, we discussed the many different forms of data teachers can access in their classroom, from quantitative to qualitative and from formally to informally generated data. Because we can easily be overwhelmed by the amount of information we collect, it is very important to remember how what we want to examine is connected within the Feedback Loop: What is the goal we are trying to get at with our data? How is the tool giving us the data we want to examine? What are the types of inferences we might make from the data? Asking these types of questions can help us be more deliberate about what we collect and reduce the amount needed for the next step—making inferences about what your students know and are able to do. Try using Resource Activity 4.1 (p. 85) to help you think about the data you want to collect and how it fits into your feedback loop.

Is It Heat? Is It Temperature?

Stephan Graham, Arrupe Jesuit High School

My name is Stephan Graham, and I am a science teacher who has worked in urban high schools for the past seventeen years in Chicago, Illinois, and Denver, Colorado. I serve a population of students from underserved and economically disadvantaged communities. The science curriculum I offer these at-promise[*] students attempts to connect challenging science topics to their everyday experiences and allows me to measure what students are learning at any given time. To that end, the four vertexes of the Feedback Loop resonated with me as an experienced teacher, as I could easily see how I already take these steps to plan and enact classroom activities daily. A good example comes from a recent lesson I taught, which focused on the difference between heat and temperature.

My *goal* for a several-day sequence of lessons was for students to determine the difference between heat and temperature. This set of lessons was part of a unit on thermochemistry, in which students ultimately create a cooking show. This is the third of four units in the chemistry curriculum in the junior year of Arrupe Jesuit, where the topics engage students with ideas that are familiar to them and align closely with the disciplinary core ideas found in the *Next Generation Science Standards* on energy, its conservation, and its transfer. In addition, concepts of heat and temperature encompass common misconceptions that the lab activities address. One common student idea I drew on as a resource is the incorrect idea that objects that feel hot to the touch must be at a higher temperature than objects that do not feel hot to the touch.

I worked with a *tool* that followed the predict-explain-observe-explain format (Figure 4.6, p 82). The tool asked students to predict whether a cube of ice melts more quickly on a metal block or a plastic block. I prompted students to think about two ideas: the difference in conductivity between these two materials and whether they believed that heat flows from hot to cold or cold to hot. Students were familiar with atomic-scale drawings of crystalline versus amorphous substances and their role in explaining thermal conductivity through these materials, but they hadn't yet learned about the direction of heat flow.

---

[*] A way of describing at-risk youth that focuses on the belief that all students can succeed.

CHAPTER 4

FIGURE 4.6 The predict-explain-observe-explain tool

Name:

Chemistry: Lab #12—Mass, Temperature, Heat Energy, and Type of Substance

Focus Questions:

1. What is the difference between heat and temperature?
2. How are mass, temperature, heat energy, and the type of substance related?

Research:

1. Which ice cube will melt faster? An ice cube on plastic or an ice cube on metal?

   Prediction:

   Observation:

   Explanation:

Then I collected multiple sorts of *data*. I first asked students individually to draw their predictions in their lab notebooks through an initial model. After a few minutes, students conducted the activity themselves, placing cubes of ice on both a metal and a plastic block and subsequently recording their observations. Next, I invited students to get into groups and use a dry-erase board to revise their models on the basis of their observations. Finally, students shared out their data with the rest of the class. Thus, I had four different sources of data: students' preliminary predictions, their initial models, revised models from each group, and the explanations that students offered to the rest of the class.

I was able to make a number of *inferences* from these sources of data about whether or not my students met my original learning goal. I was able to infer through their models and explanations that many students knew the difference in conductivity between metals and plastics, but the class as a whole had a more difficult time agreeing on the direction of heat flow. That is, they were not clear about whether heat flows from the block to the ice cube (Figure 4.7a, student A) or from the ice cube to the block (Figure 4.7b, student B). Furthermore, some students questioned if the direction of heat flow changed depending on the material from which the block was made (e.g., student B). As students listened to each other, I asked them to write down what they thought made sense. One student talked about the fact that her ice cube trays at home are made of plastic. Another student shared that for the ice cubes to melt on either the metal or the plastic block, heat must be moving from the block to the ice. A student then noticed that the metal block felt cold to the touch and offered the statement that heat must be moving from his (warm) hand to the (cold) metal block for his hand to feel cold. At this point, students felt confident that the suggestions given by the class helped explain why the ice cube melted more quickly on the metal than on the plastic block.

Collecting Data

FIGURE 4.7 Sample student models: (a) student A; (b) student B

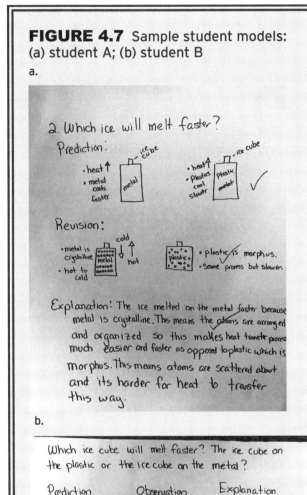

After engaging in this process of making inferences from the multiple sources of data and having students reflect on their own learning, I finished the class period with one final tool, a clicker question intended to help me get a sense of how well students had moved toward the correct explanation through the activity:

Which is at a higher temperature inside a car on a hot summer day?

a. The cloth seat covers

b. The metal buckle of the seat belt

c. Both the cloth seat covers and metal buckle of the seat belt are at the same temperature.

The data were not comforting. Students drew on their everyday experiences to answer the question; because the metal feels hot to the touch but the cloth seat cover does not, over 70% of students responded that the metal seat belt buckle would be at a higher temperature. But something interesting then happened. Only when students saw the right answer did they make a connection between the previous ice cube activity and the clicker question. Students were then quick to explain that the heat was flowing from the (warm) metal seat buckle to the (cooler) hand, just as heat flowed from the (warmer) metal block to the (cooler) ice cube. I inferred that this question also gave students an opportunity to see that heat was not the same as temperature. Everything inside the car was at the same temperature, but only the conducting materials efficiently transmitted the heat, thereby making them "feel" hotter.

CHAPTER 4

> Putting a format to the way I informally assess students is affirming. Evaluating students from the start to the end of the lesson and using the data to instruct my next lesson just makes sense. Students should show what they know periodically inside a class period and not only after a summative test.

RESOURCE ACTIVITY 4.1
Data Collection Plan

Before collecting data in your class, we suggest you take a few moments and envision how you will do so.

ASPECTS	CONSIDERATIONS	MY PLAN
Goal	What is your goal?	
Tool	Which tool(s) are you selecting, adapting, or designing?	
Class	Describe the class in which you will generate data.	
Timing	When will the data be collected?	
Categories of data (circle the ones you're collecting)	Qualitative Quantitative Formal Informal	
Instructional strategies	How are students involved in generating data? What efforts are you making to challenge and support all students?	

CHAPTER 4

References

Ainsworth, L., and D. Viegut. 2006. *Common formative assessment: How to connect standards-based instruction and assessment*. Thousand Oaks, CA: Corwin Press.

Butler, R., and M. Nisan. 1986. Effects of no feedback, task-related comments, and grades on intrinsic motivation and performance. *Journal of Educational Psychology* 78 (3): 210–216.

Chesboro, R. 2006. Using interactive notebooks for inquiry-based science. *Science Scope* 28 (4): 30–34.

Cowie, B., and B. Bell. 1999. A model of formative assessment in science education. *Assessment in Education: Principles, Policy & Practice* 6 (1): 101–116.

Duschl, R. A., and D. H. Gitomer. 1997. Strategies and challenges to changing the focus of assessment and instruction in science classrooms. *Educational Assessment* 4 (1): 37–73.

Engle, R. A., and F. R. Conant. 2002. Guiding principles for fostering productive disciplinary engagement: Explaining an emergent argument in a community of learners classroom. *Cognition and Instruction* 20 (4): 399–483.

Furtak, E. M. 2006. The problem with answers: An exploration of guided scientific inquiry teaching. *Science Education* 90 (3): 453–467.

Miles, M. B., and A. M. Huberman. 1994. *Qualitative data analysis*. 2nd ed. Thousand Oaks, CA: Sage Publications.

Rich, M. *New York Times*. 2015. Some Schools Embrace Demands For Education Data. May 11. *www.nytimes.com/2015/05/12/us/school-districts-embrace-business-model-of-data-collection.html*

Ruiz-Primo, M. A., and E. M. Furtak. 2006. Informal formative assessment and scientific inquiry: Exploring teachers' practices and student learning. *Educational Assessment* (3 and 4): 237–263.

Ruiz-Primo, M. A., and E. M. Furtak. 2007. Exploring teachers' informal formative assessment practices and students' understanding in the context of scientific inquiry. *Journal of Research in Science Teaching* 44 (1): 57–84.

CHAPTER 5

Making Inferences

It seems like I'm always in the enactment and collect data spot.

—*Experienced middle school science teacher*

The middle school teacher above, like many science teachers, finds himself collecting a lot of data every day in his classroom. However, when he had a chance to reflect on his planning process with the Feedback Loop, he realized that much of his time was spent in selecting tools and collecting data—what he called the right side of the loop—rather than looking backward at the goals that guided his work or forward at the inferences that might be made from them.

This final step of the Feedback Loop is where everything comes together, when you compare the data you collected using your tools to make *inferences* about what students know relative to the goals you originally established. We close the loop with inferences because they formally allow you to capitalize on the alignment of the goal, tool, and data, not only to look at what individual students know and are able to do, but also to look at patterns across students' responses and determine feedback to both individual students and your entire class.

In this chapter, we'll start by talking about what we mean by inferences, and then we'll walk you through a few guides for making those inferences. We'll also include some vignettes of teachers making inferences to guide their instruction.

Why Inferences?

Scientists don't stop after they collect data. They dedicate time and care to determine what those data mean. Similarly, the step of making inferences about data in the Feedback Loop

CHAPTER 5

is essential to making meaning about what students know and are able to do. The literature notes that the assessment process is founded on making "reasonable inferences about what students know" (Pellegrino, Chudowsky, and Glaser 2001, p. 42). Bennett (2011) similarly argued that formative assessment is, at its core, an inferential process, with each piece of data presenting an opportunity for teachers to refine their conjectures about what students know and are able to do. Black and Wiliam (2009, p. 13) said this is process of the teacher gaining "insight into the mental life that lies behind the student's utterances."

We have adopted the word *inference* from the assessment literature because it is very broad, encompassing both the qualitative judgments and quantitative analyses of the data you've collected. Many of the teachers with whom we've worked refer to this process as "looking for patterns" in data or "interpreting evidence" they've collected. Despite the different names, what is important to emphasize is the value of pausing and looking back through the entire framework, determining what you've learned, and deciding how this knowledge will inform your instruction (a step we'll describe in much more detail in Chapter 6).

Teachers often make inferences about their teaching and students' learning. We often hear statements like, "the kids didn't get it" or "that worked really well; possibly one of the best things I've ever done." And while it is valuable that teachers engage in these kinds of reflections, staying at that level of generality may not be particularly helpful for specifically understanding how our teaching is supporting student learning. Therefore, it is important that we ground our inferences about learning in data we've collected from our students. As science teachers, we continuously encourage students to support their claims about the natural world with evidence; in the Feedback Loop, we hold ourselves to the same standard.

Grounding Inferences in the Feedback Loop

Before we talk about processes for making inferences, we want to begin by situating those inferences in each step of the Feedback Loop: inferences follow goals, tools, and data . As a result, working through these three preceding elements helps guide and ground inferences. The advantage of the loop, then, is that it provides a structure that frames and streamlines inference-making as the teacher moves through and reflects on the different elements. The following sections will highlight the relationship of inferences to each step in the Feedback Loop.

Grounding Inferences in Goals

The whole point of the Feedback Loop is to make inferences about the data we have collected to determine if students have met learning goals. While we can make many different inferences on the basis of the data we've collected using specific tools, it is important that we remind ourselves that each inference in our feedback loop should connect to our initial goal. There are many outcomes that can be reached from data generated from students' engagement with a tool; however, the particular inferences we make through this process should be ones that inform us about our goal.

If you place your goal in a sequence, you can use the learning progression to help you identify patterns in student ideas; for example, are there responses that seem to be based on common partial understandings or preconceptions? In this way, being knowledgeable about common student misconceptions when identifying your learning goals, either because of your teaching experience or because of research, can give you an advantage in interpreting students' ideas.

Grounding Inferences Relative to the Tool

The inferences we make are affected greatly by the design of our tool. Although our tool is the only element of the Feedback Loop that's not directly connected to inferences, it serves as a direct connection between our goal and the data. As we noted in Chapter 3, tools are developed with a specific goal in mind and are meant to provide us with data that will enable us to make inferences about our goal. Weaker tools (i.e., ones that do not fit well with the goal or lead to data that provide less insight into what students know or are able to do) make it more challenging to make inferences about student learning. Similarly, if our goals are not clear, we will have difficulty making inferences that relate directly to any kind of instructionally actionable conclusion.

When you get to the point of making inferences, it's common to realize that the tool you used did not generate the data you really needed to make inferences relative to your goal (like Mr. Carlson in Chapter 3, pp. 47–52). However, sometimes just slightly rephrasing a question or adding a different one can be all it takes to help us make better inferences the next time we revisit the tool.

Grounding Inferences in Data

Claims we make about what students know and are able to do should be grounded in data. While one form of data generated by a tool aligned with your original goal will work, the more sources you can draw on, the better. Furthermore, the more varied the forms of data you use, the stronger your inferences. This could be responses during discussions, lab reports, problem sets, or anything else students do or produce in your class. Each inference you make should connect to data, and data should be used to support your claim.

Two Types of Inferences

We see the Feedback Loop as playing a vital role in making two types of inferences: everyday inferences, which are ones made on the fly to guide instruction in the moment, and deliberate inferences, which guide instruction but are made outside of class.

CHAPTER 5

Everyday Inferences

We acknowledged in the previous chapter that the data we collect about student learning may be formal or informal. Since we're constantly surrounded by informal data in our classrooms, we are also continuously making inferences. Sadler (1989) called this process "fuzzy" because it is dependent on what a teacher deems worth paying attention to and responding to at a particular time. These inferences are essential because they guide instruction in the moment and are associated with student learning (Ruiz-Primo and Furtak 2006, 2007). Teachers can make inferences on the basis of a comment a student has made and they can also collect multiple sources of information.

A room of 30 students doing a lab or engaging in a class discussion is filled with all sorts of data about students, and as a teacher, you have to decide which sources to attend to. The idea of "noticing," from the mathematics education literature (e.g., Sherin, Jacobs, and Philipp 2010; van Es and Sherin 2009), gets at this idea of picking out student ideas from the complex landscape of a classroom.

We saw in Chapters 2, 3, and 4 how you can use the Feedback Loop to help you focus on a specific goal and deliberately design tools to surface student thinking in classroom data. This process can help prepare you to make inferences on the basis of those data, since you've already thought through the types of ideas you are likely to reveal, and how you might take them up. We also know that the more sources of data you can attend to, the richer the quality of your inferences will be.

Enrique Suarez (Suarez and Otero 2014) provides an example of the importance of making on-the-fly inferences on the basis of multiple sources of data about student thinking. Mr. Suarez had the goal of engaging his students in an investigation about pitch and sound and had his students sitting in a circle around a string instrument. One student, Gustavo*, approached the instrument and said, "I know what it sounds like." He proceeded to pluck the three strings and said, "It sounds like 'ting-ting, tang-tang, tong-tong,'" as his intonation imitated the sound.

There are a number of ways Mr. Suarez could have responded to Gustavo's description of the sound. On the basis of Gustavo's intonation, Mr. Suarez already inferred that the student was noticing that different strings made distinct sounds. Mr. Suarez could have substituted academic vocabulary for sound and pitch as he was repeating Gustavo's idea; however, he instead took action to collect more data before making an inference, asking Gustavo, "Can you repeat that?" Gustavo reached for instrument, but Mr. Suarez redirected him, "Just with your mouth." Gustavo repeated his description, but this time, as he said, "Ting-ting, tang-tang, tong-tong," he used his hand, torso, and legs to further illustrate his meaning. As he said "ting-ting" with a high pitch, he held his body high, then hunched down a bit lower for "tang-tang," and even lower for "tong-tong." For each sound, he also moved his fingers as though possibly plucking imaginary strings.

Making Inferences

The canonical science being taught here was about relating how the production and characteristics (e.g., pitch) are related to some physical properties of objects (which is the foundation for understanding how vibrating matter produces waves and the propagation of waves through different media). Students had previously done an experiment in which they flicked a ruler at the end of a table, changed the ruler's length, and recorded observations.

At this point, after only seconds, Mr. Suarez used a few guiding questions and the presence of the string instrument to elicit a rich response from Gustavo. From the perspective of the Feedback Loop, his *goal* was to engage students in the phenomenon of sound production and the concept of pitch as a characteristic of sound. He used the classroom strategies, or *tools*, of engaging students in a discussion about the string instrument and asking probing questions and gathered three sources of *data* about what Gustavo knew: his words, his intonation, and his body position as representations of the sound he was hearing.

Mr. Suarez was able to use these three sources of data to make a number of everyday *inferences* about what Gustavo knew. First, because of the words and tones being spoken, Mr. Suarez suspected that the student recognized that different strings produced different the sounds. Secondly, he was able to immediately infer that the student was making a connection between the length of a string and the sound that the string produces. Finally, by looking at Gustavo's body language, Mr. Suarez saw that positioning and voice created some kind of spectrum for describing higher- and lower-pitched sounds.

By making these everyday inferences from multiple sources of data, Mr. Suarez uncovered the reasoning behind Gustavo's ideas and elevated them for other students to realize their value. Ultimately, Mr. Suarez's students took up Gustavo's "ting, tang, tong" language as they continued exploring what other physical features of the string instrument (e.g., tension on the string, frequency of vibration) affected the characteristics of the sounds produced.

If Mr. Suarez had been expecting students to use more formal, scientific language to describe the sound (such as pitch), he could have easily discarded Gustavo's description. Taking this traditional view of conceptual understanding and scientific discourse would have missed out on the substance of Gustavo's ideas, and the chance to genuinely engage with his ideas would have been lost (Coffey et al. 2011).

Mr. Suarez's treatment of Gustavo's response also reflected a view of the student's ideas as resources to guide instruction (Hammer et al. 2005) rather than "misconceptions" that needed to be confronted and replaced. As a teacher once put it in one of Erin's projects, Mr. Suarez viewed Gustavo as a glass that was "half-full" and used the ideas present to inform teaching instead of focusing on Gustavo's glass as being "half-empty."

When viewing these everyday descriptions of sound as resources, we can see how Mr. Suarez already built on them to work toward the goal of learning about sound production, in the context of an elementary science lesson. As mentioned above, those ideas could be further extended with the use of the learning progressions in the *Next Generation Science*

CHAPTER 5

Standards (*NGSS*; NGSS Lead States 2013) that illustrate how the disciplinary core idea of light and sound waves could be built across units and grade bands. As such, the everyday inferences such as those Mr. Suarez made are vitally important to teaching, and arguably the best way to listen and respond to student thinking in the course of instruction.

Deliberate Inferences

If the tool you used has a written component or has generated some other form of student data that is not ephemeral (like the student ideas shared in discussions above) but is something you can look at after class, we suggest using a more planned or deliberate process of making inferences. This is not to imply that everyday inferences are not also deliberate, but they are not made in same the slow and intentional manner.

We've worked out a sequence for guiding the process of making inferences that consists of the following seven steps:

Loop refresher: Take a moment to look back at the goal and tool you so thoughtfully developed. If you used a learning progression as part of your goal, be sure you have it nearby. If you designed a data collection guide to use alongside your tool (Resource Activity 4.1, p. 85), pull it out. Remind yourself of what you were hoping to find out (goal), as well as where in your tool each piece of that goal will be assessed.

Triage: Ideally, the Feedback Loop has limited the amount of data you collected so that it's not overwhelming. That said, we have all been in a place where we end up creating way more data than intended. If this is the case, spend a few minutes considering which sources of data you have collected will best, most accurately, and most quickly help you understand if students met the learning goals. This might involve looking at only one or two of the questions you designed or at only one or two classes of data to begin with.

Quick pass: Start by making a quick pass through the data you've collected. If you have open-ended questions, skim through student responses. If you asked a multiple-choice question, tally student responses. If students drew models, jot a few notes about what you see as you look through them.

Analysis: Once you have gotten a sense of what you're seeing, consider the distribution of different types of responses in the data you've collected. What groups do you see? What patterns emerge? These initial hunches are like the first drafts of your inferences, which you then check against a larger amount of data or by going through the data again to be sure you didn't miss anything.

Going back to the data: Now take those hunches and go back to the data a little more thoroughly. If you only looked at part of your data, look at all of it this time; if you breezed through everything the first time, go through it again a little more slowly. On this pass, you should see if there are data that confirm the hunches you identified or if you can catch anything that goes against them (what researchers would call "disconfirming data" (Miles and Huberman 1994). If you find that the majority of the data align with your initial thoughts, you're done and have a finished set of inferences. If you find enough cases that conflict with your ideas—that is, enough disconfirming data—record the pattern you observed in the disconfirming data as an inference.

Summarize your inferences: Once you've been through this process, write down the final inferences you've made about student thinking; Resource Activity 5.1 (p. 101) can guide you through this process. These inferences will serve as your final guide for giving students feedback. We encourage you to start this process of summarizing by noting, "What do students know now relative to the original goal? What do they know that I can build on tomorrow, even if it's technically 'wrong'? How might I organize students in groups so that they can draw on each other as resources?" You can include partial understandings, examples, and so on. Then, note what students are still working on. What important ideas are they confused about that you can clear up tomorrow? What activities can you build into the lessons in the next few days to help students move forward in their learning?

Look behind; look ahead: You know you won't remember later in as much detail as now what you'd like to change next time around about the prompt or your instruction that led up to the data you're analyzing. In your lesson plans from this year, on a sticky note, a digital comment, or somewhere else you'll find it next year, jot yourself a note about what you would do differently next time. At the same time, use these fresh inferences to look ahead.

Involving Students

One of the strongest strategies for involving students in assessment is to engage them in the process of peer or self-assessment (Coffey 2003; Sadler 1989). When applied to the Feedback Loop, student self-assessment fits into what we call inferences; however, rather than having you make inferences about what students know, students make these inferences themselves. Shepard (2000, p. 12) said that "student self-assessment … promises to increase students' responsibility for their own learning and to make the relationship between teachers and students more collaborative," thus helping to establish a learning culture in the classroom.

In peer assessment, students use the quality criteria they helped establish in the goals step to make inferences about what their peers are able to do. In many cases, this involves passing work back to your class, and asking students to evaluate the quality of what their peers have done and then write feedback to help students improve their work. One teacher Erin worked with adapted a listing of students' common ideas about natural selection into a handout called "Correct the Statements," and she provided students opportunities to respond to different types of student ideas and talk about how they might be improved.

In self-assessment, students evaluate their own progress and make inferences about their own standing relative to learning goals. This step, as we mentioned above, is essential in the learning processes: Ultimately, students have to change their performance for themselves. In this step, students "compare the actual (or current) level of performance with the standard, and … engage in appropriate action which leads to some closure of the gap" (Sadler 1989, p. 121). Another teacher Erin worked with provided her students with the same constructed-response tool multiple times in a unit and gave her students different colors of pens so that they could go back and modify their own responses. This helped the students be more metacognitive of their own learning processes.

Rubrics can be helpful guides for both peer and self-assessment by helping students locate themselves or their peers along a continuum of proficiency. Given this information, they could make inferences about what they know or where they might have questions and could develop ideas for next steps to address these areas of need.

Summary

Congratulations! You've gone through the four elements of the Feedback Loop. However, what remains is arguably the most important step in affecting student learning, that is, using the inferences you made to inform the next steps for your instruction. We saved this part for Chapter 6 because, as Bennett (2011, p. 14) stated, "if the inferences about students resulting from formative assessment are wrong, the basis for adjusting instruction is weakened." Now, let's build on these inferences and decide what to do to move students toward learning goals.

One Day, Multiple Feedback Loops

Deb Morrison, Broomfield Heights Middle School, Broomfield, Colorado

I am a middle school science teacher-researcher engaging my students in the process of modeling about Earth science phenomena. I center my work in questions relevant to students' lives or of interest due to my students' innate curiosity about the world. I went through multiple trips through the feedback loop in a single day as I gradually improved the quality of the tools, data, inferences, and feedback I provided students in an eighth grade plate tectonics unit. My students find earthquakes to be a fascinating and engaging topic, so I focused my whole unit on plate tectonics around this phenomenon. I started with a puzzling question, "What causes earthquakes?"

I had already done some initial assessment at the beginning of the unit that allowed me to infer that only about 5% of my students had anything approaching a scientific explanation for the source of earthquakes. Most students used the term "plate tectonics" in a vague way that did not include an understanding of plate movement. Like Wegener's detractors, my students were incredulous that the ground could be moving over an inner mantle of fluid rock. Where was the mechanism for that? Students also struggled with the scale of time over which these processes take place; the idea that time can extend for millions and billions of years when talking about Earth's history is difficult for students to understand. We had already wrestled with this idea during our erosion unit, but students had started to build an evidence-based understanding that small processes over long periods of time could transform the surface of the Earth. Finally, in terms of argumentation, my students were quite adept at constructing claims and supporting them with evidence by the plate tectonics unit; however, they were still having trouble connecting evidence to claims through the use of scientific principles—the reasoning piece.

Early in the unit, I had engaged students in a number of activities about earthquakes and the evidence gained from studying them. This included how evidence of the way in which different types of earthquake waves differentially travel through the center of the Earth has helped scientists come to understand the fluid nature of the mantle. Many of my students were surprised—and a little disturbed—to find out that the Earth they were standing on was not actually so solid after all. This new learning was connected to our prior lessons on the convection currents that occur in the atmosphere, understandings gained during our study of weather and climate. Once students accepted that the Earth's mantle could have convection currents, they began to see how plates might exist and be moving across the surface of the Earth.

CHAPTER 5

This piece of the unit was structured around the learning goals shown in Table 5.1.

TABLE 5.1 Summary of goals

Standard or performance expectation
Colorado Grade Level Expectation: *Major geologic events such as earthquakes, volcanic eruptions, mid-ocean ridges, and mountain formation are associated with plate boundaries and attributed to plate motions; Students can Identify, interpret, and explain models of plates motions on Earth.*

NGSS DIMENSION	CLASSROOM FOCUS
Science practice	*Engaging in argument from evidence:* Argue with fossil and other evidence to support a claim about plate tectonics.
Disciplinary core idea	*ESS2.B:* Plate tectonics and large scale systems interactions: Plate tectonics is the unifying theory that explains movements of rocks at Earth's surface and geological history. Maps are used to display evidence of plate movement.
Crosscutting concept	*Scale, proportion, and quantity:* Examine the different temporal and spatial scales of evidence to support plate tectonics.
Student ideas as resources	• Vague use of "plate tectonics" with no understanding of plate movement • No mechanism for plate movement • Long periods of time could change Earth's surface but deep time (millions or billions of years) is still a struggle • Claims and evidence are understood but reasoning is a challenge

I used this set of learning goals to frame the way I constructed and organized assessment activity in my classroom. My instructional goal for the class was to have students begin to reason with evidence about the theory of plate tectonics, the idea that Earth has plates moving across the mantle due to convection in the mantle. Up to this point, students had not explicitly learned about plate tectonics but rather had focused on the phenomenon of earthquakes. Students had begun to notice patterns in the location of earthquakes and to understand that the inside of the Earth was exerting a great deal of energy at particular places, which might cause the surface of the Earth to change. At the beginning of the first class devoted to this goal, I gave short, just-in-time instruction on the theory of plate tectonics, outlining that the evidence of particular fossils and plate boundary shapes in particular suggested a different historical arrangement of the continents than what we see today. I then provided my students with the materials from the USGS Wegener's Puzzling Evidence Exercise (*http://volcanoes.usgs.gov/about/edu/dynamicplanet/wegener*) and asked them to use the continental puzzle pieces provided to create a single supercontinent that may have existed.

During this lesson, I had several opportunities to make inferences about student learning goals. Figure 5.1 illustrates the lesson sequence and the locations of different assessment moments using particular tools.

Making Inferences

FIGURE 5.1 Classroom activity and assessment

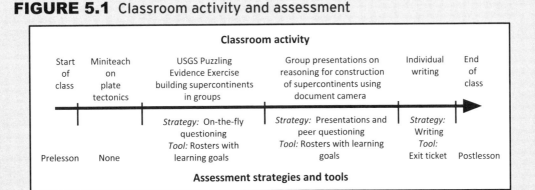

The first *tool* I used was improvised questioning as students worked in groups constructing their supercontinents. I moved systematically through the room during this 15-minute period, stopping to ask the spokesperson for each group (defined roles that I rotate to ensure I hear all students) questions such as, "Why are you putting your pieces together this way?" or "Why do you think the evidence supports this arrangement of pieces?" The students' responses to these questions served as my *data*; sometimes students would not provide reasoning, leading to my in-the-moment *inference* that students needed support in providing connections between their claims and evidence. In such cases, I provided quick feedback in the form of a follow-up question such as, "Why does it matter if a fossil on one continent matches up with a fossil on another continent? Couldn't those organisms just evolve on different continents?" These questions pressed students' reasoning about the connection between evolution and fossil evidence. As I circulated, I recorded student responses on my classroom roster, noting who I heard from (Figure 5.2) and my on-the-fly inferences about their responses.

FIGURE 5.2 Rosters with learning goals for a sample of students

Student	Mantle movement	Claim stated	Evidence reference	Evidence reasoning
Juan Rodriguez*	−		+	
Kylee Munger*			+	+
Sean Schreter*	+	+	−	
Joanna Orduna*	*	§	+	−

Key: − = not scientifically accurate; + = scientifically accurate but not in scientific language; § = scientifically accurate and in scientific language. No marking indicates that student did not say anything about this.

CHAPTER 5

The second feedback loop built on the same set of goals, and I used the tool of the students' written activity again to collect another source of data by having students share their reasoning with each other. Groups came to the front of class, shared their supercontinent images under the document camera, and multiple members of each group talked through their reasoning as to why they organized their puzzle pieces as they had. Peers then asked questions about their reasoning. In each instance, I was able to collect data from both the presenters' comments and the peers' questioning, which helped me make more inferences about their thinking. If I was unclear at any point of their meaning, I would interject with a question such as, "What do you mean by the edges of the continent fit? In what way did they fit?" These questions were simultaneously opportunities for me to get more data about what students were thinking and ways to act on my inferences in the moment. This helped students be specific in the way that they were referencing evidence, such as continental boundary shapes versus contiguous fossil finds. Again I recorded student thinking on my roster.

The final feedback loop I completed during this class period, the exit ticket, involved writing and allowed me to collect data from every student to guide a more systematic set of inferences. I had the students fill out the exit ticket silently and individually after they had time to process their thinking through peer-to-peer talk. This exit ticket, shown in Figure 5.3, had a little more space for students to write complete sentences.

FIGURE 5.3 Exit ticket for the plate tectonics activity

Exit Ticket—Plate Tectonics Evidence   Name: _____
                                            Period: ___

In your own words, describe the theory of plate tectonics.

What evidence have you explored that supports the theory of plate tectonics? **Provide at least two pieces of evidence and explain why they support the theory of plate tectonics.**

1.

2.

Rather than take the exit tickets home, I checked them as students handed them in, sorting them into piles according to my learning goals: no clear understanding or reasoning of plate tectonics, partial understanding of plate tectonics or some evidence and reasoning illustrated, or scientific understanding of plate tectonics and evidence and reasoning illustrated. At this point, I could clearly see that although most of my students had gained a partial understanding and could provide some evidence to support their theory, most were still struggling to understand what reasoning was and how to engage scientific principles such as convection in the mantle or evolution of an organism to plate movement or fossil evidence. To close the feedback loop, I centered the next two days' work on more detailed examinations of evidence and reasoning around evidence.

I put all of these tools in place as the day progressed. I teach five sections of Earth science, and between classes I took the inferences I had made in the prior class and applied them to the next class so that my lessons got better as the day went on, due to my ongoing collection of data about student learning. The several feedback loops I went through during the day are shown in Figure 5.4 (p. 100).

During my first class, I enacted the USGS activity exactly as described in their materials. However, I found that the students appeared to rely on the shape of continental coast lines more heavily than fossil evidence to support their claims about continental movement, and I was unable to understand students thinking on an individual level. In addition, I wanted students to engage in more complex reasoning with fossil evidence. The information gained in this initial feedback loop (0) prompted me to develop an exit ticket for individual formative assessment, which allowed me to gain a better understanding about why students were drawing on particular evidence to support their claims. In the next feedback loop (1), I was able to gain much more detailed information about student thinking and found that students were not always using details in evidence and were having a lot of difficulty reasoning with scientific principles. Thus in the final feedback loop of the day (2), I revised the USGS activity by pre-preparing puzzle pieces to minimize off-task time and maximize opportunities for classroom talk of evidence and reasoning. This seemingly small adjustment made a big difference in the amount and quality of student talk because students immediately began to use the information they had to argue with each other about why pieces should go in particular places. As a result, the students' responses to the exit ticket in my final class of the day were the most targeted. My inferences throughout this day led me to design a two-day lesson following this class that began with a miniteach on Earth's internal structure and convection currents and then focusing on reasoning with a variety of types of evidence about plate movement.

CHAPTER 5

FIGURE 5.4 The multiple feedback loops used throughout the day of teaching about plate tectonics

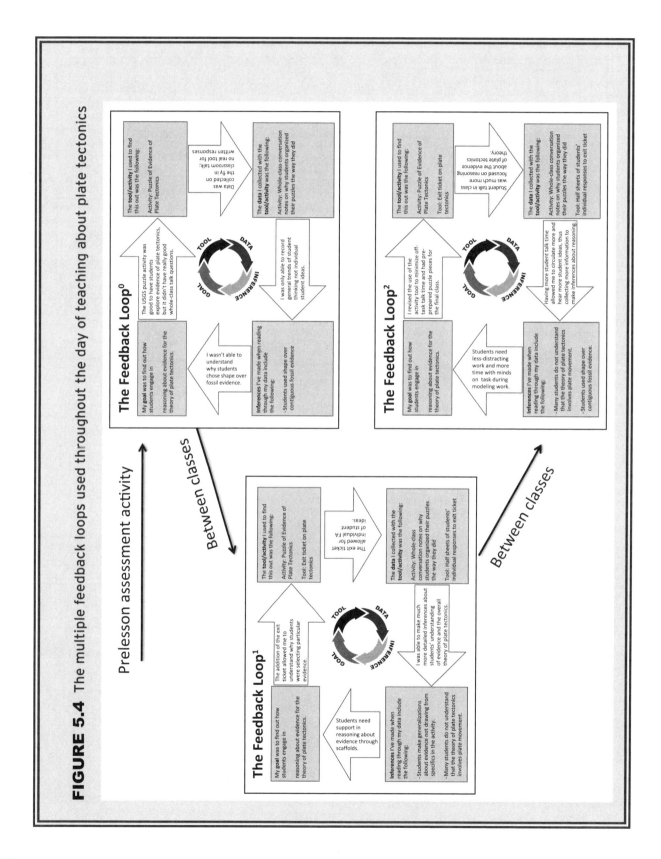

Making Inferences

RESOURCE ACTIVITY 5.1
Guide to Tracking Inferences

My goal was

The tool I developed was

Everyday or common ideas I hoped to surface included (resources for my instruction)

My sources of data included

My inferences include

THINGS THE STUDENTS UNDERSTAND THAT I CAN USE AS A RESOURCE FOR MY INSTRUCTION	WHAT THE STUDENTS NEED TO WORK ON

THE FEEDBACK LOOP Using Formative Assessment Data for Science Teaching and Learning

CHAPTER 5

References

Bennett, R. E. 2011. Formative assessment: A critical review. *Assessment in Education: Principles, Policy and Practice* 18 (1): 5–25.

Black, P., and D. Wiliam. 2009. Developing the theory of formative assessment. *Educational Assessment, Evaluation and Accountability* 21 (1): 5–31.

Coffey, J. 2003. Involving students in assessment. In *Everyday assessment in the science classroom*, ed. J. M. Atkin and J. E. Coffey, 75–87. Arlington, VA: NSTA Press.

Coffey, J. E., D. Hammer, D. M. Levin, and T. Grant. 2011. The missing disciplinary substance of formative assessment. *Journal of Research in Science Teaching* 48 (10): 1109–1136.

Hammer, D., A. Elby, R. E. Scherr, E. F. and Redish. 2005. Resources, framing, and transfer. In *Transfer of learning: Research and perspectives*,. ed. J. Mestre, 89–120. Greenwich, CT: Information Age Publishing.

Miles, M. B., and A. M. Huberman. 1994. *Qualitative data analysis.* 2nd ed. Thousand Oaks, CA: Sage Publications.

NGSS Lead States. 2013. *Next generation science standards: For states, by states.* Washington, DC: National Academies Press. www.nextgenscience.org/next-generation-science-standards.

Pellegrino, J. W., N. Chudowsky, and R. Glaser. 2001. *Knowing what students know: The science and design of educational assessment.* Washington, DC: National Academies Press.

Ruiz-Primo, M. A., and E. M. Furtak. 2006. Informal formative assessment and scientific inquiry: Exploring teachers' practices and student learning. *Educational Assessment* (3 and 4): 237–263.

Ruiz-Primo, M. A., and E. M. Furtak. 2007. Exploring teachers' informal formative assessment practices and students' understanding in the context of scientific Inquiry. *Journal of Research in Science Teaching* 44 (1): 57–84.

Sadler, D. R. 1989. Formative assessment and the design of instructional systems. *Instructional Science* 18: 119–144.

Shepard, L. A. 2000. The role of assessment in a learning culture. *Educational Researcher* 29 (7): 4–14.

Sherin, M., V. Jacobs, and R. Philipp, eds. 2010. *Mathematics teacher noticing: Seeing through teachers' eyes.* New York: Routledge.

Suarez, E., and Otero, V. 2014. Leveraging the cultural practices of science for making classroom discourse available for emerging bilingual students. In *Learning and becoming in practice: The International Conference of the Learning Sciences (ICLS) 2014, Volume 2*, ed. J. L. Polman, E. A. Kyza, D. K. O'Neill, I. Tabak, W. R. Penuel, A. S. Jurow, K. O'Connor, T. Lee, and L. D'Amico, 800–807. Boulder, CO: International Society of the Learning Sciences.

van Es, E. A., and M. G. Sherin. 2009. The influence of video clubs on teachers' thinking and practice. *Journal of Mathematics Teacher Education* 13 (2): 155–176.

CHAPTER 6

Closing the Feedback Loop

Feedback means hearing them. It means, especially when we're in a discussion, affirming all of their ideas ... hearing what they're saying and maybe rephrasing it in more ... words that are science words instead of maybe more colloquial words. It also means when I make a mistake or they make a mistake, this didn't work, how come? Maybe we didn't do this correctly. Maybe we need to look at this again. Feedback also means acknowledging what's not working but in a way that doesn't threaten them as people.

—Alice Schafer*, middle school science teacher

This book is about a process for planning and reflecting on formative assessments in secondary science classrooms, and we call it a loop for a reason: Once you've gone through each of the four main steps in the process, the idea is to connect the inferences you've made back to the goals. This process of closing the loop is often called feedback in the formative assessment literature (Black and Wiliam 1998) or, put more simply, using the information you've gained in your trip through the loop to move students forward in their learning (Ruiz-Primo and Furtak 2006, 2007; Shepard 2000).

While the term *feedback* can sound a little abstract, we like to think about it as something that effective teachers use every day when they are being responsive to information about student thinking. But, research into feedback has clearly identified approaches to responding to student thinking that are more and less effective at influencing student learning (Hattie and Timperley 2007).

In this chapter, we will first talk about the process of connecting inferences about student learning back to your original goals in the feedback loop, a process that we call identifying

CHAPTER 6

the gap. Then, we will introduce a number of different categories of feedback. Finally, we will share a vignette of an experienced middle school science teacher and use transcripts from a whole-class discussion to illustrate the variety of ways she provides feedback to her students.

Formative Assessment and the Feedback Loop

Let's get this chapter started by connecting the process that we have described as the Feedback Loop with common ways of describing formative assessment.

Formative assessment is commonly characterized as a three-step process in which a teacher sets learning goals, determines what students currently know, and then provides feedback to support students in meeting those goals (NRC 2001). Other authors have written about an important, intermediate step in this process, which is comparing the information about what students currently know to the learning goals to determine the size of the gap between what students know and are able to do and what is expected (Bennett 2011).

The Feedback Loop overlaps with these descriptions: We start by setting learning goals, and then the next three steps—designing and selecting tools, collecting data, and making inferences—are a way of explicating the "determining what students currently know" step. The piece that is not represented as a stand-alone step in the Feedback Loop is the final element of formative assessment, which is providing helpful feedback to move students toward learning goals. This step is the way we think about that final arrow connecting inferences and learning goals. It's the feedback that connects what you have inferred about what students know and are able to do with the goals you originally had for student learning.

At the end of the day, many authors have argued that it's this last step in the Feedback Loop that is most essential to help students learn (Black and Wiliam 1998). The literature on how individuals use feedback to improve their performance is pretty conclusive: Although you might provide people with information about their performance, at the end of the day, individuals are the ones who take action on feedback to improve their performance (Sadler 1989). When applied to teaching, this means that students need to have their own understanding of quality that is relatively similar to the teacher's to regulate and improve their performance. Tweed (2009) summarized research in involving students in the assessment process and noted that it

> *builds relationships among students and sets expectations that students will support one another's learning and use habits of mind—such as seeking accuracy and clarity, keeping an open mind, responding appropriately to others' feelings and level of knowledge—that make students feel accepted and capable in the classroom. (p. 180)*

Although the word "feedback" may suggest an immediate response in the moment to student thinking, Wiliam (2007) has talked about multiple types and lengths of feedback loops that connect what you learned with adjustments to your teaching and feedback for student learning. This means that the feedback you take to "close the loop" might happen within a few seconds of surfacing a student idea, but it might also happen later in the lesson. If you collect students' written work, you might take the evening to look it over and provide feedback the next day. Further, that information you gleaned about student thinking could inform what you do the next week or even in the next unit of instruction. Finally, you can consider the information as a way to fine-tune your instructional activities for the next time you teach the unit, with another group of students. Keep these multiple time frames in your head as you read the following sections about types of feedback. Resource Activity 6.1 (p. 119) summarizes these different lengths of feedback loops.

Evaluative Versus Informational Feedback

There are a lot of ways of thinking about feedback for students, but not all are equally effective. In fact, in a meta-analysis that looked across studies of the effects of feedback on student learning, Hattie and Timperley (2007) found that not all methods of giving feedback to students were helpful in promoting student learning.

Evaluative feedback: The most common way teachers respond to students is by telling them whether they are right or wrong. It's easy to understand doing this. When a student has shared her thinking, it can sound crazy or outright wrong, and as teachers who know science, we're definitely tempted to tell students that they missed the mark. Otero and Nathan (2006) called this a "get it or don't" way of viewing learning: Either students get it and are therefore told they are correct, or they don't get it, in which case they must be told they are wrong.

Unfortunately, responding to students this way in a formative assessment setting is a double-edged sword. First, by evaluatively judging students once they have shared their ideas, we are in essence shutting down their thinking. Students are not likely to be motivated to tell us what they really think if we tell them immediately they are right or wrong. Doing so plays into the game students know well by secondary school—that teachers are there to tell them if they are right or wrong (Furtak 2006).

Second, evaluating right off the bat cuts off chances to further explore student thinking. We have found in the course of our own teaching and in our work with teachers that often student ideas that initially seem wrong or convoluted might be naive versions of the "right" answer just phrased in terms we don't recognize, or students might blend together scientific terms with everyday meanings of words. Examples of this include students talking about objects that are dense as being thick (Shemwell and Furtak 2010); thinking that fitness in evolution is the same as being in good shape rather than focusing on reproductive success; or talking about vehicles speeding up rather than accelerating.

CHAPTER 6

Learning theory pushes us to consider students as having a range of ideas rather than sets of right or wrong ideas that need to be replaced. This way of thinking about student learning is better thought of as a continuum that connects students' everyday experiences in their lives with the scientific ideas and practices we expect them to learn in school. Thinking back to representing this process as a feedback loop means we want to connect the experiences brought out in the data we collect about student learning with the goals we originally set. Making this process continuous with what students know is definitely better than cutting off what we found out through the feedback loop and just saying "you're right" or "you're wrong."

Finally, results of research are pretty clear on evaluative feedback. It is not as effective at helping students learn as other strategies (Hattie and Timperley 2007), which we will discuss in more detail next.

Informational feedback: Rather than giving students feedback that is evaluative, research suggests that providing them with information based on what you know about them, that is, the inferences you have made about what they know and are able to do on the basis of the data you collected, leads to higher learning gains (Hattie and Timperley 2007). Informational feedback provides some kind of information that describes not only the differences between the students' current performance and learning goals but also about how to get there.

Dylan Wiliam (2007) illustrated this type of informational feedback by describing how a coach can observe the motion of a softball pitcher and ultimately provide feedback that helps her improve her performance:

> *Consider a young fast-pitch softballer who has an earned-run-average of 10 (for readers who know nothing about softball, that is not good). ... Analysis of what she is doing shows that she is trying to pitch a rising fastball (i.e., one that actually rises as it gets near the plate, due to the back-spin applied), but that this ball is not rising and therefore becomes an ordinary fastball in the middle of the strike zone, which is very easy for the batter to hit ... her rising fastball is not rising, which is why she is giving up a lot of runs. (p. 1,062)*

So we have a pitcher that has the goal of throwing a rising fastball, and the coach asked her to demonstrate her motion (in a way, this is the task or tool), which led to the coach making a series of inferences about what is happening. In effect, the coach is identifying the gap between the desired goal and what is actually occurring, or the fact that the rising fastball is not rising but rather going down the center of the strike zone. The final step to close the loop, then, is to provide feedback that will result in an improvement of the pitch. In Wiliam's scenario, the pitching coach tells the pitcher that "she is not dropping her shoulder sufficiently to allow her to deliver the pitch from below the knee" (Wiliam 2007, p. 1,063). This feedback doesn't just tell the pitcher what she already knows—that her fastball is not rising—but contains specific information about how to improve. This makes the coach's action formative and helps shape learning.

It is important to point out that the reason the feedback was so helpful to the pitcher was because the coach knew not only what the best motion for a rising fastball looks like but also, most likely, the most common ways that pitchers can make mistakes that lead to the fastball *not* rising. This ability to make inferences on the basis of observational data about performance is central to providing effective feedback. Bennett (2011, p. 17) described it as the ability to

> *consider the distinctions among errors, slips, misconceptions, and lack of understanding. An error is what we observe students to make—some difference between a desired response and what a student provides. The error we observe may have one of several underlying causes. Among other things, it could be a slip—that is, a careless procedural mistake; or a misconception, some persistent conceptual or procedural confusion (or naive view); or a lack of understanding in the form of a missing bit of conceptual or procedural knowledge, without any persistent misconception. Each of these causes implies a different instructional action, from minimal feedback (for the slip), to re-teaching (for the lack of understanding), to the significant investment required to engineer a deeper cognitive shift (for the misconception).*

When we take this example into a science classroom, it becomes readily apparent why feedback is so challenging for teachers to enact—more challenging, in fact, than any of the other elements of formative assessment or the Feedback Loop (e.g., Heritage et al. 2009). To give informational feedback, science teachers need to thoroughly understand what they are teaching, and they must also be able to identify and distinguish among the "errors, slips, misconceptions, and lack of understanding" (Bennett, 2011 p. 17) of science concepts and practices. Without these warranted inferences, teachers will struggle to determine how to give feedback to students.

However, once teachers have thoroughly come to understand the goals they are teaching, have designed tools that will surface student ideas relative to those goals, and have collected data, the inferences become easier to make because they are aligned with what was expected; the feedback can almost be planned out in advance. A teacher in one of Erin's studies spent months going through this process of unpacking the goals and designing a tool, and on the day that she was giving the assessment she had codesigned, Erin was at the back of the room with a video camera to record what she was doing. The first time a student shared one of the misconceptions the assessment was designed to elicit, the teacher walked close to the camera, stared straight into the lens, and mimed incredible excitement before straightening her face and going back to teaching. Because she already anticipated the student's response, she already had her next steps in her pocket!

A Few Feedback Strategies

The tough thing is that, when it comes to feedback, there is no easy prescription of what to say or do because every piece should, by definition, be uniquely tailored and responsive to

CHAPTER 6

the inferences you've made about learners. Coffey and colleagues (2011) talk about feedback as having disciplinary substance, or information in it that is scientific, which attends to the nature of student thinking and practices. That said, the types of informational and helpful feedback you give students can be thought of as fitting into a few larger classroom teaching strategies or formats where you are in a mode specifically geared to close the feedback loop.

Cuing or pushing talk moves: In the chapter about data, we talked about assessment conversations as one way of making student ideas explicit in whole-class discussion. This strategy can be an efficient way of getting information about what students know and are able to do because it can be done in real-time; it also creates opportunities for teachers to get further information about what students know through follow-up questions.

At the same time, these assessment conversations can be important opportunities for teachers and students to engage together in feedback that moves students toward learning goals. A number of researchers have identified this talk pattern as cuing (e.g., Kluger and deNisi 1996). A teacher in one of Erin's research studies once used the metaphor of first letting sheep out to pasture and then herding them into the pen. The first step involves letting students out to be free to share ideas and wander around in them. However, at the end of the day, the teacher is accountable for helping students toward the right answer. This is where cuing moves come in. Mortimer and Scott (2003) talked about this as alternating different types of patterns in the classroom, first engaging in dialogue to get students to air out their ideas (or wander the pasture), and then being more "authoritative," to use their language, to move students toward certain ideas.

Feedback in an Assessment Conversation

Rachel Tanner*, a high school biology teacher, had just engaged her class of ninth graders in a multiple-choice assessment related to the causes of adaptations in *Biston betularia* (peppered moths). Ms. Tanner had codeveloped this assessment in collaboration with the other biology teachers in her department around the *goal* of determining if students were able to identify the origin of variations within a population.

The *tool* the teachers developed took advantage of a multiple-choice plus justification format so that students read a short story about *B. betularia* during the industrial revolution and were then asked to select one of four responses to a multiple-choice item as follows:

Closing the Feedback Loop

The difference in the number of light- and dark-colored peppered moths in the population can be explained by

A. *birds, which were more likely to eat the light-colored moths than the dark-colored moths. The darker moths were then more likely than the lighter moths to survive and reproduce.*

B. *industrial pollution, which caused the bark of trees to darken over time. This caused the coloring of moths to become darker over time.*

C. *a change in the color of bark, which caused the genes in a moth to mutate to produce a darker color. Then the moths with the mutation of a darker color were more likely to survive.*

D. *the moths themselves, which were able to make themselves become a darker color to match the bark of the trees, which had been darkened over time by industrial pollution.*

Why did you choose your answer?

_____

Ms. Tanner provided students about 10 minutes to read the assessment, select an answer, and then write an explanation for why they had chosen the response they did. She then collected two additional sources of *data*: students' votes and their explanations shared in a whole-class conversation.

Ms. Tanner: It sounds like the discussion is over, so let's share as a class. First of all—hold on. First of all, let's just do a quick show of hands. How many people chose D? OK—is that—how many people chose C? B? Option B? And, finally, A, I'm guessing is the remainder? Um, I think there are like twenty-three-ish.

Ms. Tanner wrote the votes on the board: 23 votes for A, 1 for B, 0 for C, and 1 for D. This quick survey let her know immediately that the majority of the class had chosen the correct answer. However, Ms. Tanner wanted to create space for students to talk about all the responses. She said afterward, "I wanted Ellie* [the student who picked D] to explain why she thought what she did, but I didn't want to make her feel singled out. … I did see some other answers from the students, and most of them picked A." She initiated a discussion of the students' justifications, first calling on Ellie.

Ms. Tanner: All right, so, let's start with Ellie. What compelled you about D? Why, what made you think that was a good answer?

Ellie: Um, 'cause it talked about a mutation, and I just thought that the mutation made it more evident that it was from a change in the environment. We were

CHAPTER 6

	saying that mutations [inaudible], but it was rather because they had adapted to their environment.
Ms. Tanner:	OK. So, when it said the moths themselves, which were able to make themselves a darker color to match the bark of the trees, which have been darkened over time by industrial pollution. So you think that the moths changed color because of the pollution?
Ellie:	Sure.
Ms. Tanner:	OK. Um, B.
Jessie*:	That's exactly what I chose. I chose B.

At this point, Ms. Tanner had collected some information about Ellie's reason for choosing response D, and Jessie shared that he also chose another answer for that same reason. However, Ms. Tanner didn't want Ellie to feel singled out, noting, "It's hard when they are learning in front of everybody." She chose to close the feedback loop by asking leading questions of her students that would help Ellie—and anyone else with her same answer justification—to see the inconsistencies in the response.

Ms. Tanner:	Because B and D are actually very similar, right? OK. So, do you guys remember, you guys remember this? So, it was this idea that in the U.S. we have lots of resistant lice, right, to things like RID? And the two ideas are, the two ideas are that in a given population of lice you have resistant versus nonresistant lice, or that exposure to the lice shampoo actually caused a mutation, right? And which one was it that we figured out?
Students:	A.
Ms. Tanner:	Hypothesis A. OK. So let's go back to D. Option D. It says, the moths themselves, which were able to make themselves become a darker color to match the bark of the trees, basically resulted because of the pollution. Do you see any similarities there?
Student:	Lamarck?
Ms. Tanner:	Exactly. So, this is hard, it's a really subtle difference, because for so long, you guys have learned that, you know, the organisms change, adapt to fit their environment, and they do, but that's not a—that's the end result. Do you know what I mean? It's not the mechanism by how that happened. And so, um, it's—it's confusing, definitely.

Here Ms. Tanner provided some informational feedback, where she related students' experiences in a lab activity back to the response options. She then opened the conversation again for explanations for choosing response option A.

Ms. Tanner:	So, people who picked A, why did you choose A? Let's hear from some, yes. Thank you.
Lisa*:	Well, because, um. I just lost my train of thought. Oh, because the light ones were easier to see by the birds, so the birds ate more of those, which left more

	of the darker population of lice, or not lice, moths to reproduce, which therefore made the population greater of dark moths, because, there was more.
Ms. Tanner:	It changes the gene pool, right? Exactly. Any other reasons why people picked A?

After surfacing reasons for response option A, Ms. Tanner transitioned to cuing feedback, now taking students through a chain of reasoning that related several classroom activities to the idea that she wanted to help them understand at the end of the lesson.

Ms. Tanner:	Again, this whole idea that your environment changes you is not necessarily the case, right? The environment, again, think about the predator activity we just did. Did you do anything to those dots?
Students:	No.
Ms. Tanner:	No, but how did the population change by the environment? Did they mutate in any way?
Students:	No. Ones that were more fit.
Ms. Tanner:	So the variations existed in the population to begin with. Right? Do you think change like that can occur if variation wasn't there in the population?
Students:	No.
Ms. Tanner:	No, right?

Ms. Tanner then looped back to Ellie to see if she was able to clarify her understanding.

Ms. Tanner:	So, is this, Ellie, is this clear? I mean let me know if there's something else, because people in seventh period also said, "Well, it's the environment that causes the change," and that's partly true, but natural selection doesn't happen—mutations and changes in populations don't happen because you're exposed to a certain environment. The environment you're in sort of exacerbates or brings out the different, um, the different gene—genotypes that would do better in that particular environment. But it's not like it can create it if it wasn't there to begin with. So, birds don't get wings because they needed to fly. What happened, do you think?
Students:	There were some that had wings, and some that didn't.
Ms. Tanner:	There were some that had some sort of form of webbing, possibly, right? And that helped them to get from one place to another better. And what did they do with those genes? They passed them on, so on and so forth. Right? It's not that they needed to fly and that's why they got wings.

This exchange, while short, illustrates Ms. Tanner starting with a more open style of questioning and then drawing purposefully on in-class examples to help students reinterpret the response options. Reflecting back on this experience, Ms. Tanner said the following:

CHAPTER 6

> *I think [Ellie] gets it, but it would be interesting to have her do just one more thing to see if her notions really have changed. So what I was planning on doing was giving them back the individual change or natural selection sheet and having them revise their answers with what they know now. So I was going to do that maybe in the next two periods or so. So I think I'll probably do that on Friday. So that way they can see, "Oh gosh, wow. What was I thinking?"*
>
> Here Ms. Tanner noted that although she made the *inference* that Ellie had "gotten it," she was still not sure, and she planned an additional feedback activity for later in the week, in which she would give all of her students back an activity and allow them to reflect on their prior responses, revising them if they chose to do so.

Whole-class redirect: We've all been there: You get a lab going, students start collecting data, and you quickly realize the reagent you provided them with is the wrong molarity, so everyone's data is coming out wrong. What do you do? Many of us go to the front of the room, ask everyone to pause, and then identify what's going wrong so that students can continue with their work.

This everyday strategy, which we'll call a redirect, takes only a few seconds, but can also work really well for feedback that attends to your inferences about student ideas and practices. As you're going around the class looking at data about student learning and making inferences on the fly, you usually notice patterns in what students are doing. For example, students writing explanations may be defending claims but not supporting them with evidence (Berland and McNeill 2010). A quick redirect to the whole class can help you provide informational feedback to everyone at once:

> *Hey, everyone, can you look up here for a moment? As I've been going around the class, I noticed that many of you are making claims, but you are not providing evidence to support those claims. Be sure that your explanations include claims, evidence, and reasoning.*

This simple redirect has taken only a few seconds, but it contains a wealth of information for students to improve their performance. You've shared your inference with them, that they are not providing evidence. You also share information to help them improve their performance, that they need to be sure to include evidence and reasoning along with their claims.

Another version of the whole-class redirect can happen in the next class period. Imagine that you had students complete an activity (tool), and you spent the afternoon that same day going through their responses (data). As you go through their responses, you note a few patterns (inferences) in what they wrote and how they responded, so you jot a few notes into a slide and show it to them at the start of class the next day.

Closing the Feedback Loop

Yesterday at the end of class, I handed everyone an exit ticket that gave you a chance to explain to me, in your own words and using pictures, the process of diffusion. I took these home with me last night and looked them over, and I noticed several things that I wanted to revisit before we move on to the next lesson today.

Again, this redirect has provided students with a reminder of the goal, the tool you provided them, and your inferences about patterns you observed in the data you collected. One possibility to redirect the class would be to talk about each of the patterns and then suggest what those ideas are missing; another possibility would be to engage students in the conversation and have them talk about the ideas. We'll see a more in-depth example of this at the end of the chapter.

Reteaching: During Erin's dissertation research, she observed two different teachers guiding students through an investigation that compared the displaced volume and mass of sinking and floating objects. The distinction between displaced and total volume was tricky for both the teachers and students. Both teachers encountered difficulties in having students graph their data. The trend line, which should have had a roughly 1:1 slope, could not even be drawn because the data points were all over the graph. Both teachers realized the mistake: Students had measured total volume instead of displaced volume.

The important difference is in what came next. One teacher merely substituted a graph from a different class period and showed it to students, stating that this was how their data should have come out. The other teacher, however, realized her mistake and took some time to have students remeasure their data and then make the graph again. This teacher, rather than plowing ahead, created time to provide information directly about what had gone wrong and gave students the opportunity to remeasure—and ultimately relearn—what they needed to know. In contrast, the other teacher skimmed over the mistake, and his students came away more confused.

We know what you are thinking. The curriculum is already so crammed; there is no time for reteaching. We acknowledge that this is an extreme example, and it is, of course, unrealistic to reteach every lesson in which students struggle with understanding. However, we also know that for those really important concepts that are, in essence, the linchpins of key understandings, taking the extra time to support students in developing foundational knowledge and practices will ultimately serve their learning going forward.

Summary

In this chapter, we talked about how informative feedback is essential to effective formative feedback. Specifically, the three strategies of cuing, redirecting, and reteaching are just a few among a plethora of ways you can structure your class to provide feedback and close the loop. Perhaps the best mantra to keep in mind, one that we draw from the mathematics education literature, is that your teaching should allow you opportunities to notice what

CHAPTER 6

your students are thinking and to attend to that thinking in the course of your instruction (e.g., Sherin, Jacobs, and Philipp 2010). The Feedback Loop helps you formalize this process by taking you through a set of steps that prepare you for the final arrow, which closes that gap. The vignette that ends this chapter illustrates how one teacher provided feedback in a variety of ways to support student learning.

One Tool, Two Sources of Data

Alice Schafer*, the experienced middle school science teacher you read about in Chapter 3, was helping her students develop evidence-based explanations for sinking and floating through measuring and collecting data, and searching for patterns in sinking and floating objects. She had engaged her students in a series of investigations in which they made observations and took measurements of mass and volume of sinking and floating objects, and made graphs of mass versus volume to identify patterns in objects that sank or floated. Ms. Schafer then set the *goal* of determining if her students were able to use the evidence from the investigations in which they had engaged to make an explanation for why things sink and float. To find that out, she used the *tool* shown in Figure 6.1, which was developed as an embedded assessment in the curriculum.

FIGURE 6.1 Ms. Schafer's sinking and floating tool

Name_____ Teacher_____ Period___ Date_____
Please answer the following question. Write as much information as you need to explain your answer. Use evidence and examples to support your explanation. **Why do things sink and float?**

Source: Ayala et al. 2008.

This is where Ms. Schafer got creative. She used this single tool to collect multiple forms of *data* about what her students knew and were able to do, and within these sources of data, embedded sets of *inferences* that she was collecting about student thinking.

The students took about 10 minutes to respond to the question in their science notebooks at the end of class, and then Ms. Schafer spent about 30 minutes reading through her 17 students' written responses after class. She immediately noticed several patterns in students' ideas, which constituted her first set of inferences about student thinking. She noted that some students were drawing on evidence in their explanations and others were not. Several were making unsupported claims about sinking and floating. Furthermore, some students still had ideas that were associated with common misunderstandings, and others were using everyday language (e.g., using "thickness" to describe density). These inferences she made about student ideas were based on a quick read-through of the students' work.

She found a few pieces of colored paper and wrote the summarized ideas onto them. "Because different colors make different ideas stand out, I think I'm putting together things that somehow, in my mind, go together as a way of ... getting them to see that things relate." The ideas on the piece of paper read as follows:

1. The further the mass is to the displaced volume, it will sink.

2. Size, shape, and weight are related.

3. Material is mass.

4. Shape, size, and volume are related.

5. Water is more dense than air and oil. It has more matter, so it has more mass.

6. Density is how thick it is.

7. Thickness of a liquid

8. If an object is bigger, the [blank] mass is spread out, and it will float.

9. The more mass, the more sinking only works if two objects have the exact same volume.

The next day in class, Ms. Schafer used magnets to put the different ideas on the pieces of paper on the board one at a time and asked students to share their thinking about the ideas. In this way, she collected another set of data about their thinking by listening to them talk about the different ideas.

Ms. Schafer set up the classroom conversation by summarizing what she wanted to do.

Ms. Schafer: And I really, I was impressed with how hard you worked. I was really impressed with how much evidence you got. I've never seen you quote so much evidence. It was awesome. So I'm going to put some of your ideas up here on the board, and we're going to figure out which ones we want to keep and which ones we figure aren't the real reasons. So, now we're looking for the reason [why things

CHAPTER 6

sink or float], OK? And, so, Jack said—well, let's not claim these ideas, because the thing is that we may decide not to keep one, and I don't want you to take it personally.

She then called on students to talk about the different ideas on the pieces of paper, and students shared their evidence around the different ideas. An example exchange is illustrated below.

Ms. Schafer: If an object is bigger, the blank mass is spread out, and it will float. Now, this one confused me, and that's why I put a blank here. An object is bigger—so we have the small carton, the large carton, and we have four grams of matter—oh, half-full.

Then, Ms. Schafer saw the ideas on the board as an opportunity engage students in a conversation to clarify what they meant by those ideas. For example, she asked students to elaborate on idea 8.

Ms. Schafer: So, we have the same mass? We took the same mass from the little carton, and what did we do with it, Henry?
Henry*: We poured it in the big carton.
Ms. Schafer: Poured the same mass in the big carton, and it didn't sink as far. So, you're going after this idea of spreading out, but if the object is bigger, the same mass is spread out? Is that what you mean? The same amount of mass? If an object is bigger, the same amount of mass is spread out, and it will float.

Ms. Schafer had spent some of her class period allowing students to talk about the ideas shared and clarifying what they thought. The first set of ideas she wrote on the board represented some inferences about patterns of ideas she had seen, and the class discussion led her to make an additional set of inferences about what students thought. She then moved to providing students feedback to move them closer to the ultimate goal by clustering similar ideas together to help them see similarities and differences in the ideas.

Ms. Schafer: The more mass, the more sinking, only if the two objects have the exact same volume. OK, now, did we see any words here—do you see any things here that go together? That's the same ideas that we should put next to each other? Kenny, what do you think?
Kenny*: [Shape, volume, and size.]
Ms. Schafer: Shape, size, and volume. You think those are—how many people agree that these belong together? Well, because this one says size, shape, and weight. … Is weight a science word we use?
Students: No.
Ms. Schafer: What is the word we've been using?
Students: Mass.
Ms. Schafer: Mass. So do we want to keep this one?

Students: No.

Ms. Schafer: OK, we don't use the word weight. We haven't used the word at all. We always mass things on a balance. So, do these three things have a similar idea to them? The shape of something and the size of it and the volume? Should we choose one of those words as the science word we're going to use? What is your opinion? Which one can we measure? We measured one of these, and the others we haven't really measured. What do you think?

Ms. Schafer also used the posting of the students' ideas as an opportunity to invite disagreement around the ideas.

Ms. Schafer: Do you agree with this one? If the object is bigger, the same mass is spread out. It will float. If you have a bigger object, and you take the same mass, it's going to float?

Laura*: Yeah.

Ms. Schafer: And I'm going to put this over here by mass. What are we going to do with this density? How thick it is? Density is how thick it is.

Laura: With the water one, it's hard to do a backflip in the air, but you can do it in water more easily.

Ms. Schafer: That's interesting. So, something about the water is helping you do a backflip. Can you do it as fast as in the air?

Students: No.

Ms. Schafer: No, but can you still do it?

Students: Yeah.

Ms. Schafer: So, there's something about the water you're saying that supports you? Kind of supports you? Does it have this thick idea or not? When you think about doing a backflip in it, does it have this thick density idea?

Laura: Yeah.

Ms. Schafer: You want to keep that? Shall we keep that up here with this one?

In this exchange, Ms. Schafer actively helps her students connect a naive idea about thickness to the scientific concept of density. She uses guided questions and student examples to help make these connections. In so doing, she is pushing students to see the relationships among different ideas and helping them to move forward in their thinking.

Ultimately, Ms. Schafer used one goal and tool to generate two sources of data and sets of inferences: students' written responses and students' shared ideas in a whole-class discussion as data, and her inferences about both of those. How Ms. Schafer's actions fit into the Feedback Loop is summarized below. Something remarkable about this process is that Ms. Schafer collected these data without committing herself to large amounts of time outside of class. She did not grade students' work but simply flipped through it after school to get a general sense

CHAPTER 6

of what they had written. After recording these general impressions on paper, she held a class discussion, which presented another opportunity for students to surface their thinking, and which also served the purpose of giving students feedback about their ideas. Figure 6.2 summarizes Ms. Schafer's lesson in a feedback loop.

FIGURE 6.2 Sinking and floating in the feedback loop

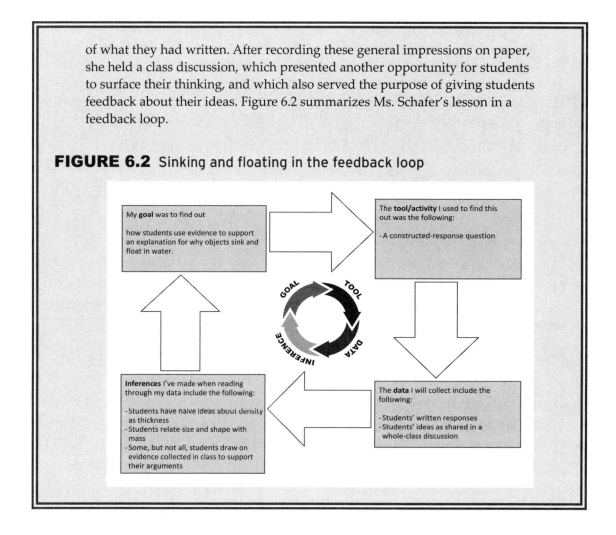

Closing the Feedback Loop

RESOURCE ACTIVITY 6.1
Multiple Feedback Loops

Now that you have spent time analyzing your data, we'd like to push you to consider how you might use your findings and interpretations of the data to inform your teaching.

Take a moment to reflect on other ways you can "close the loop" using your inferences
- if you could go back five seconds after you had collected the data?
- if you could go back to five minutes after you had collected the data?
- looking ahead to the next class period?
- looking ahead to next week?
- when you move into the next unit this year?
- when you teach this lesson again next year?

CHAPTER 6

References

Ayala, C. C., R. J. Shavelson, M. A. Ruiz-Primo, P. Brandon, Y. Yin, Y., E. M. Furtak, and M. Tomita. 2008. From formal embedded assessments to reflective lessons: The development of formative assessment suites. *Applied Measurement in Education* 21 (4): 315–334.

Bennett, R. E. 2011. Formative assessment: A critical review. *Assessment in Education: Principles, Policy and Practice* 18 (1): 5–25.

Berland, L. K., and K. L. McNeill. 2010. A learning progression for scientific argumentation: Understanding student work and designing supportive instructional contexts. *Science Education* 94 (5): 765–793.

Black, P., and D. Wiliam. 1998. Inside the black box: Raising standards through classroom assessment. *Phi Delta Kappan* 80 (2): 139–148.

Coffey, J. E., D. Hammer, D. M. Levin, and T. Grant. 2011. The missing disciplinary substance of formative assessment. *Journal of Research in Science Teaching* 48 (10): 1109–1136.

Furtak, E. M. 2006. The problem with answers: An exploration of guided scientific inquiry teaching. *Science Education* 90 (3): 453–467.

Hattie, J., and Timperley, H. 2007. The power of feedback. *Review of Educational Research* 77 (1): 81–112.

Heritage, M., J. Kim, T. Vendlinski, and J. Herman. 2009. From evidence to action: A seamless process in formative assessment? *Educational Measurement: Issues and Practice* 28 (3): 24–31.

Kluger, A. N., and A. DeNisi. 1996. The effects of feedback interventions on performance: A historical review, a meta-analysis, and a preliminary feedback intervention theory. *Psychological Bulletin* 119: 254–284.

Mortimer, E. F., and P. Scott. 2003. *Meaning making in secondary science classrooms*. Berkshire, UK: Open University Press.

National Research Council (NRC). 2001. *Classroom assessment and the National Science Education Standards*. Washington, DC: National Academies Press.

Otero, V., and M. J. Nathan. 2008. Preservice elementary teachers' views of their students' prior knowledge of science. *Journal of Research in Science Teaching* 45 (4): 497–523.

Ruiz-Primo, M. A., and E. M. Furtak, 2006. Informal formative assessment and scientific inquiry: Exploring teachers' practices and student learning. *Educational Assessment* 11 (3 and 4): 237–263.

Ruiz-Primo, M. A., and E. M. Furtak. 2007. Exploring teachers' informal formative assessment practices and students' understanding in the context of scientific inquiry. *Journal of Research in Science Teaching* 44 (1): 57–84.

Sadler, D. R. 1989. Formative assessment and the design of instructional systems. *Instructional Science* 18: 119–144.

Shemwell, J. T., and E. M. Furtak. 2010. Science classroom discussion as scientific argumentation: A study of conceptually rich (and poor) student talk. *Educational Assessment* 15 (3 and 4): 222–250.

Shepard, L. A. 2000. The role of assessment in a learning culture. *Educational Researcher* 29 (7): 4–14.

Sherin, M., Jacobs, V., and Philipp, R., eds. 2010. *Mathematics teacher noticing: Seeing through teachers' eyes*. New York: Routledge.

Tweed, A. 2009. *Designing effective science instruction: What works in science classrooms*. Arlington, VA: NSTA Press.

Wiliam, D. 2007. Keeping learning on track: Classroom assessment and the regulation of learning. In *Second handbook of mathematics teaching and learning*, ed. J. F. K. Lester, 1053–1098. Greenwich, CT: Information Age Publishing.

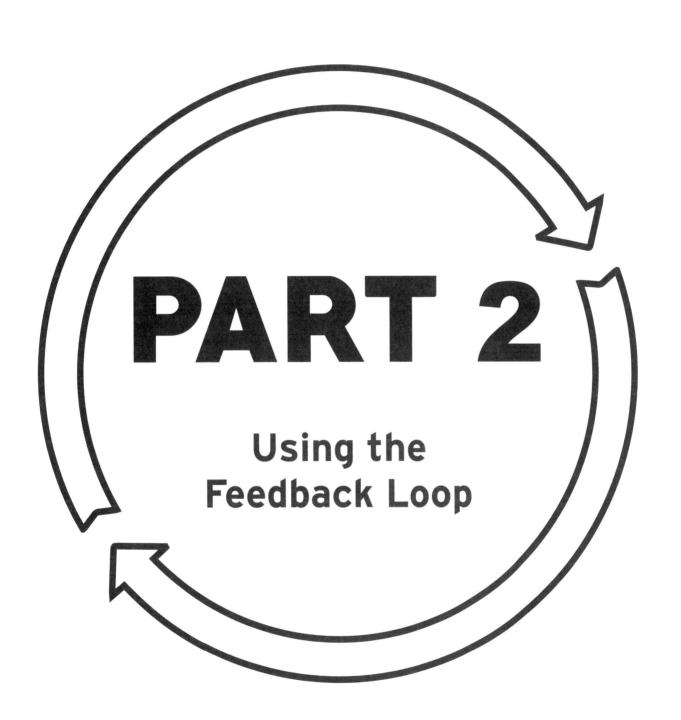

CHAPTER 7

Using the Feedback Loop to Plan and Inform Instruction

The idea that you're sort of slowly improving on your practice makes it a lot more manageable—instead of saying you're going to come out of the gate your first year with perfect lesson plans every day.... Just that iterative process of going through the feedback loop is really helpful.

—Erin Zekis, Arrupe Jesuit High School

At the outset of this book, we hoped to connect with what is a common experience for practicing secondary science teachers; that is, to be overwhelmed by the large amount of data about teaching and learning that is generated daily. In the preceding chapters, we have illustrated that part of feeling this way comes from thinking about data on its own and not seeing its connections to the other three elements of the Feedback Loop.

Now that we have dedicated a large portion of this book to discussing each of the elements of the Feedback Loop—setting goals, selecting tools, collecting data, making inferences, and providing feedback—we will bring all of the elements together. Our intention is to illustrate how thinking within the Feedback Loop can help you set specific, measurable goals, design tools for collecting data, guide your interpretation of it, and then determine next steps for instruction.

This chapter is designed to combine all the information you've seen in the previous five chapters into a series of resources that will help you learn to use all the elements of the Feedback Loop together. Our hope is that these exercises will help you think about the interrelationships among the elements.

CHAPTER 7

Reflecting on Previously Collected Data

In our experience working with teachers as they learn to use the Feedback Loop, we have found that it's constructive to start by looking at a set of data they collected in a previous unit or semester. We all have these stacks of papers—tests, quizzes, worksheets, lab reports, science notebooks—that we keep putting off again and again. Why might that be? Are there simply too many questions to grade? Are you not sure where to start? Which of the items on the test or quiz or which pages in the notebook should you look at first? How will you ever find the time to give every student feedback on everything they have produced? Situating your first trip through the loop in data you already have helps bring this way of thinking about goals, tools, data, and inferences into focus in the context of your own work. A version of the feedback loop that can guide you through this process is shown in Figure 7.1, and Resource Activity 7.1 (p. 137) will also guide you through each of the steps described below.

FIGURE 7.1 Reflecting on a previously collected set of data

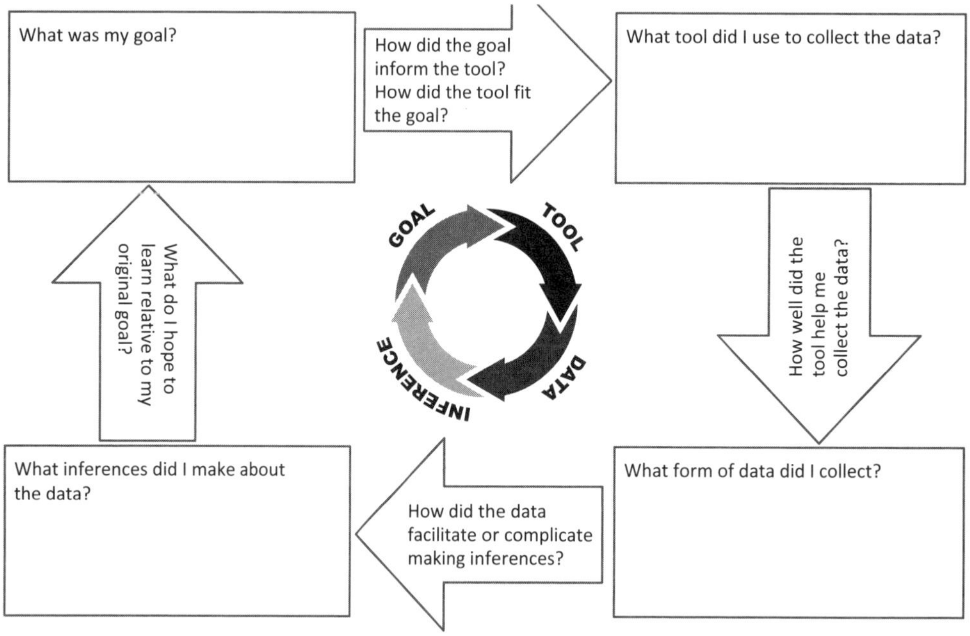

First, take a step back and be honest with yourself as you consider why you collected this data in the first place (i.e., think back to your original goal). Was the data specific and measurable? If you are in a state that has adopted the *Next Generation Science Standards* (*NGSS*), did it feature multiple components in a performance expectation? If your original goal doesn't meet these criteria, how might you revise it?

These reconsiderations should lead you into a process of questioning whether the tool you used to collect the data you're looking at aligned with your original goal. Take a moment

Using the Feedback Loop to Plan and Inform Instruction

to read through the tool and consider its structure and each question and item in light of your revised learning goal. Which questions are really getting at what you want to know? Which ones are peripheral and might you be able to eliminate or ignore?

Looking at the data again after considering the goal and tool can help you see through the multiple questions that students might have responded to and hone in on what's important. What did students actually write in response to the questions or items you determined were most important? You might notice that instead of having 15 questions per student, now you only have 2 or 3, immediately restricting the data you really need to examine to determine what students have learned. You have now used the first two elements of the framework, goals and tools, to help you quickly cut through all of that information to what is most important.

Finally, take a moment to reflect on inferences you might have initially made from these data, if any. If you never looked at them—as all three of us have been guilty of in many instances—this might not apply. However, if you did, what inferences did you make about individual students and your class as a whole? In what ways were these aligned with the goal you originally had in mind? It's quite possible that in looking back at the data, you will realize that the inferences you originally made were not necessarily aligned with the goal you had in mind or with the revised, more focused version of that goal you have now identified. This is a common occurrence, since feeling overwhelmed with data can lead to grading approaches that focus more on surface features of student responses (e.g., Which vocabulary terms did students use? Did they label both their axes and title their graph? Did they write in complete sentences?) rather than the deep understandings they have that are related to the goal. Looking just at the shorter subset of questions, however, might lead you in a different, more focused direction that is better aligned with your original goal.

Once you've finished going through this process, take a few steps back to consider takeaways from this process that you might be able to apply beyond this set of data. Were your goals too broad? Was the tool too long, or did it yield too much data? Were your inferences about students aligned with what you intended to learn? These reflections will help guide your future data collection as, in many cases, they help to identify themes in classroom practice that can yield insights for your future work with data about student learning. Going through this process just once can yield such insights, and going through it more than once can help you identify trends in your teaching even better.

Planning for Instruction With the Feedback Loop

Usually, going through this retrospective process of data analysis illuminates a lot about how you're currently setting goals, selecting tools, collecting data, and making inferences on the basis of those data to inform your next instructional steps. If you feel like the different elements were not in alignment, don't worry—we are now asking you to use what

CHAPTER 7

you've learned about how the elements relate to each other to make a forward-looking plan. Figure 7.2 and Resource Activity 7.2 (p. 139) present a version of the Feedback Loop very similar to what you've seen previously. The important difference is that instead of looking back on what you've done before, these versions project into the future to guide your planning, instruction, and reflection on the data you collect. As you complete each section of the framework, we encourage you to refer back to Chapters 2–5 and the specific suggestions they provide about each element. If you're doing this alongside your colleagues, Chapter 8 will provide a process for collaboratively developing, enacting, and reflecting on common assessments.

FIGURE 7.2 Making a plan to collect data

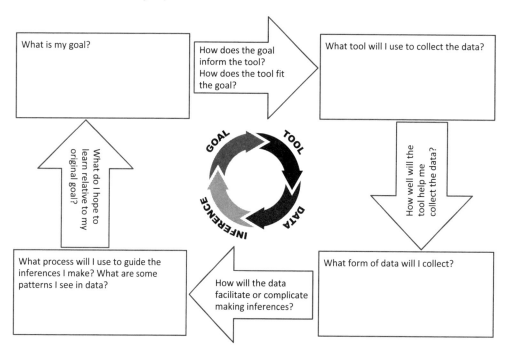

The Feedback Loop in Action: Renee's Classroom

Renee Simon* is a high school biology teacher who has been working with a university-based research team for several years to support her in working with her colleagues to design and enact assessments for her natural selection unit. Renee and her colleagues decided that they wanted to know how well students were able to develop well-articulated explanations for natural selection using a list of provided terms.

With this goal in mind, they developed a handout that would guide students to use particular words in developing a concept map about natural selection. Although Renee was interested in what the concept maps looked like, her prior experience had led her to

Using the Feedback Loop to Plan and Inform Instruction

understand that the conversations about the maps were often more interesting than what the students wrote down. Student small-group conversations are a classic example of informal, qualitative data generated by a group work tool; however, Renee wanted to have a more formal way of capturing students' qualitative descriptions of the terms, and their explanations of natural selection using those terms. She created and printed out a table that would help her record what she heard students say. These two tools (the concept map and the table) were thus created specifically to match her goal and combined to help her create formal, qualitative data about student thinking.

On the day Renee enacted her lesson, she put students into groups of three to four to create their concept maps. After students got started, she began walking around the classroom, pausing for a few minutes at each table to jot down notes about the different ideas she heard students share. Although she was primarily listening, she also asked follow-up questions in some cases. Pointing at one link on a concept map, Renee asked the group, "What does this mean? How does this affect natural selection? How do all of these concepts fit into this one? I think you guys have great connections; it's just that the big concepts are difficult to connect." Each time students responded, Renee would add to her sheet of notes.

At the end of class, Renee had collected two different forms of data relative to her goal: from the concept map tool, she now had a stack of poster-sized, group-constructed concept maps about natural selection. In addition, Renee had a page of notes about the ideas she had heard students share.

After class, Renee sat down to interpret the data she had collected. In her work with her colleagues, she had previously identified areas that might be problematic in her students' explanations and that she specifically wanted to look for as she made inferences about what students know. She had seen several different examples of learning progressions, or representations similar to those in the *NGSS*, that laid out how student thinking about natural selection develops in high school and where students often struggle in learning these concepts. For example, she knew students often did not understand how mutations led to new variations within populations of organisms. Furthermore, she knew that some students came to class thinking that organisms were able to change themselves in response to environmental changes and would often make statements using phrases like "need to adapt" or "want to adapt" in connecting different concept terms together.

A quick overview of the concept maps helped Renee see that although some of her students were representing concepts as networks of ideas that would indicate more integrated networks of understanding, such as the group shown in Figure 7.3 (p. 128), others created circles out of the concepts, linking them together in sentences without making more complex connections (see Vanides et al. 2005). Then, turning to her notes (Figure 7.4, p. 128), Renee saw that students were using language about variations and mutations that indicated understanding of the relationship between the two things.

CHAPTER 7

FIGURE 7.3 Detail from a sample concept map about natural selection

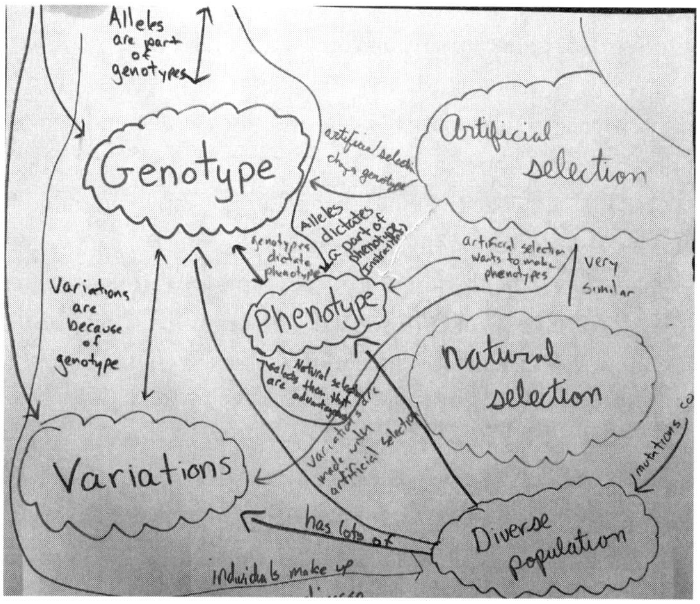

FIGURE 7.4 Renee's notes about the natural selection concept map conversations

Concept Map #1–Variation/Random Mutation
Period 1

Table #	Comments
1	• Mutations occur in an indiv. DNA • Individuals can't create mutations • Phenotype, Genotype and Alleles are more closely together than the rest (worked alone)
2	Alleles → Both have to do with genes Genotype Geno + Phenotype — both have to do with genes = no work • New generations over and over are ancestry • G.V. should explain all these words
3	• Alleles make genes • Variation makes us us indiv. • I think if we stack the words on top of each other it would make better sense • Let's define these words first (after they placed others on sheet)
4	• Mutation can screw up the variation • Variation can cause • Mutations cause variation • Individuals create generations • If you talk about change, you have to include all these terms

Bringing all this back to the goal of determining the quality of explanations students were making about natural selection, Renee considered what she had found. Although students seemed to be making explanations using terms such as variation and mutation, they still appeared to be struggling to represent the interconnectedness of multiple ideas

Using the Feedback Loop to Plan and Inform Instruction

in their concept maps. Furthermore, on the basis of what she heard students say in their small groups, Renee concluded that there was disagreement within the class as to the role that mutations played in new variations and whether those processes were random or not.

In considering next steps for her teaching and for student learning, Renee considered how to close the feedback loop on multiple levels, as we had discussed earlier. With respect to the activity, Renee jotted a few notes regarding how the actual activity itself might be reframed next year to lead to better concept maps across her entire class; for example, how the prompt for the activity might encourage students to make connections among multiple concept terms. In addition, she wondered how her students might be encouraged to share their maps with the whole class and discuss the quality of various links among concept terms. As for activities that were coming up in the next week, Renee determined that she wanted to engage students in another activity about random genetic processes that would connect more strongly to what they had learned in their genetics unit, focusing on how processes such as deletions or substitutions in base pairs, crossing-over, or recombination of genes through sexual reproduction could lead to new variations within a population.

Renee's four steps are illustrated in Figure 7.5, from starting with a goal to determining the next steps for student learning. As stated earlier, her goal of understanding how students explained natural selection was intended to assist her in learning more about what students knew about this content. She was then able to use inferences arising from this to decide what to do next with these students and how to strengthen this unit in the future.

FIGURE 7.5 Renee's reflections in the feedback loop

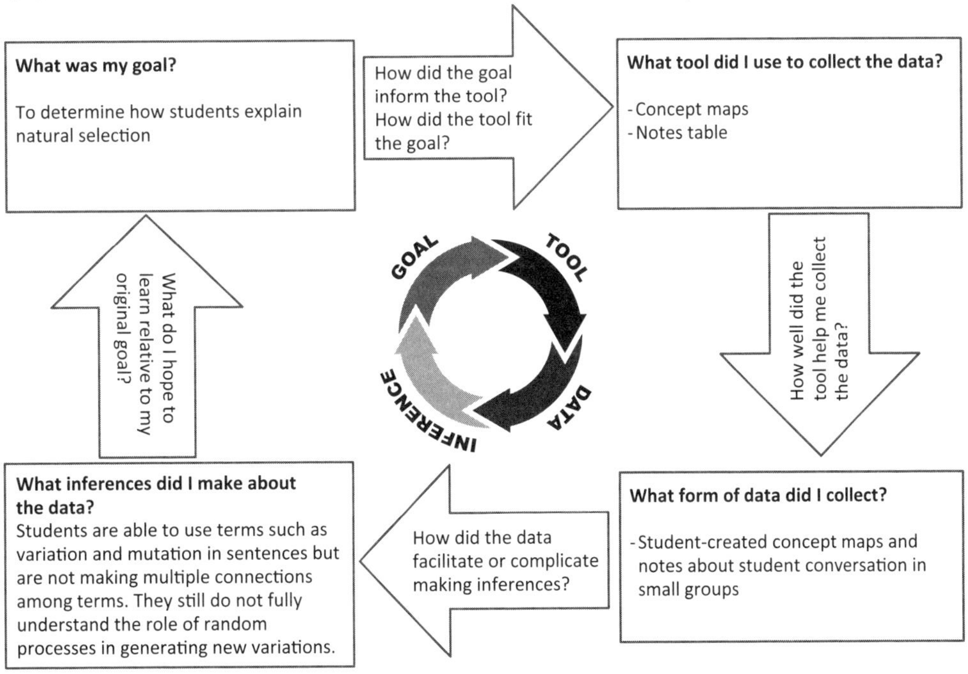

What was my goal?
To determine how students explain natural selection

How did the goal inform the tool? How did the tool fit the goal?

What tool did I use to collect the data?
- Concept maps
- Notes table

How well did the tool help me collect the data?

What form of data did I collect?
- Student-created concept maps and notes about student conversation in small groups

How did the data facilitate or complicate making inferences?

What inferences did I make about the data?
Students are able to use terms such as variation and mutation in sentences but are not making multiple connections among terms. They still do not fully understand the role of random processes in generating new variations.

What do I hope to learn relative to my original goal?

CHAPTER 7

Cascading Sets of Goals

We acknowledge that in addition to closing feedback loops with your students on multiple levels, going through the analysis framework may also lead to more goals, not only in terms of what students know and are able to do but also questions for your own teaching practice. This is where the iterative nature of the framework comes in. Your inferences lead you to identify areas in which to provide feedback *now*, but they will also lead you to more questions that will inform a new cycle of data analysis. Perhaps you have uncovered areas of prior knowledge that you want to address with students after going through the cycle or, after feeling like students were able to participate in science practices that you wanted them to, you want to move into another area. In this way, we see your initial time through the feedback loop as potentially the first of many steps in a process that can support continuous improvement of teaching and learning over long periods of time (Figure 7.6).

FIGURE 7.6 Multiple feedback loops leading to new goals

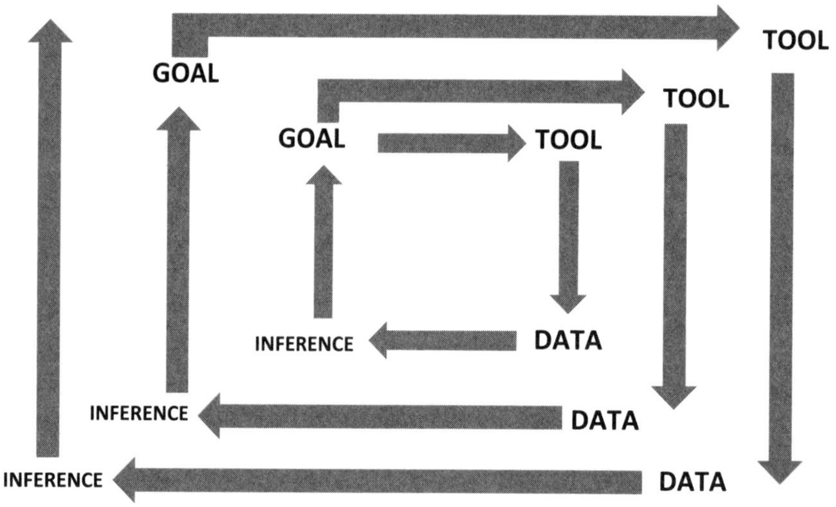

Additional Steps: Involving Students and Engaging With Colleagues

In this chapter, we have illustrated how to use the analysis framework as a whole, provided resources for reflecting on data you've collected in the past and data you could collect in the future, and suggested mechanisms for identifying types of feedback you could provide students on the basis of what you learn from completing the framework. In the next chapters, we will discuss how to engage in this process of goal identification, tool selection, data collection, and inferences with your colleagues.

Using the Feedback Loop to Plan and Inform Instruction

Refining Practice With the Feedback Loop

Erin Zekis, Arrupe Jesuit High School

My name is Erin Zekis ("Miss Zekis," occasionally "Miss Geekis," or often just "Miss"). I teach a ninth-grade conceptual physics and math seminar at Arrupe Jesuit High School. Our students are chosen from the local population partly on the criterion that they all come from low-income households. This private school has a special arrangement for which students subsidize their tuition through a corporate work-study position, which they attend five days per month. This not only helps students afford a Catholic education but also gives them work experience. As you can imagine, Arrupe Jesuit is a unique teaching environment.

I came to Arrupe as a student teacher in 2013; my cooperating teacher, Stephan Graham, also contributed to this book. Stephan has been at this for a while, whereas I'm just getting started in the profession. If you're relatively new to teaching too and are looking for a beginner's viewpoint on the Feedback Loop, I hope my reflections are helpful to you.

This year, Stephan and I participated in a digital professional development circle with other teachers in nearby communities. This Virtual Teaching and Learning Community, or vTLC, allowed each of us to practice using the Feedback Loop as we bounced ideas off each other and to hear what experiences others were having in the science classroom. This community aspect was perhaps the most important: Although I can get feedback and teaching ideas from Stephan and other colleagues at my school, I valued removing the barriers that often separate teachers from their counterparts at other schools—especially here in the United States.

Our iterative process of meeting and discussing regularly led us to a common goal and interest: models in the science classroom. Most of us agreed that, while we rely on models all the time, we rarely—if ever—consciously and explicitly discuss what models are, why they are important, what their limitations are, how to use them, and so on. We were also aware that Developing and Using Models is one of the eight science and engineering practices of the *NGSS*, with good reason: Models are ubiquitous in science, and fundamental to what science is. We decided to practice the Feedback Loop in the context of models and to share how that played out in our different content areas and classroom settings.

My ninth-grade physics classes were just beginning to plan and conduct investigations about sound. While I thought about models and sound as well

CHAPTER 7

as the lab activities and concepts I had already mapped out for students in our curriculum, I was attracted to the idea of collecting data on students' conceptions of how sound phenomena function. I was aware from some perfunctory research that students often held incorrect ideas about sound being an energy form that can permeate matter, rather than being intimately tied to matter as its medium.

I was also aware that it's a very advanced science practice to generate models on one's own. Practicing scientists work on it for a living! This is especially difficult if one's background in related disciplinary core ideas, such as the atomic and molecular makeup of matter, is not particularly strong.

I therefore considered ways to engage students in science practices around models in a more "entry-level" way. I chose a *goal*, which was to examine how well students would be able to use evidence to discriminate between two models to determine which provided a better explanation for the phenomena observed.

I started by giving students a quick *tool* that would give me some preliminary *data* on students' ideas of how sound worked so I would have a better idea of where they were coming from. I did this with an exit ticket that included two questions: (1) "What is sound?" and (2) "How do you think sound works?" I asked them to use a combination of words and pictures to express their ideas.

As I had anticipated, there was a mix of ideas. This quick trip through the feedback loop helped me make *inferences* that a small handful of students already had quite accurate and sophisticated ideas of how sound worked, describing or drawing molecules vibrating to make a wave. Most students used vague terminology and very simple drawings, indicating to me that they were operating mostly from general experience. They had heard of sound referred to as a wave but hadn't been asked to formulate how these waves behave or how they function. This allowed me to better frame the overarching feedback loop I was working on.

I made a draft of a *tool*, which was a poster with two drawings of competing sound models (Figure 7.7) distilled from the student responses I had collected in my first trip through the loop. The first model, "Sound as an Energy Wave," shows a transverse sound wave that is separate from and passing through the surrounding medium. The second, "Sound as Vibration," shows a longitudinal compression wave, with the medium's particles vibrating to propagate the sound wave.

Using the Feedback Loop to Plan and Inform Instruction

FIGURE 7.7 A tool to elicit student thinking about models of sound

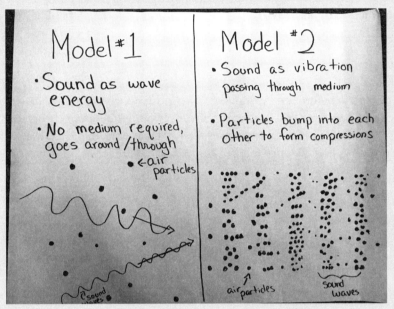

I shared these models with my teacher collaborators in my vTLC group who helped me to fine-tune wording to help align the assessment, and the resulting data, with my original goal. With their guidance, I made a table that had several rows with different video evidence. I selected some videos that clearly showed sound as vibration and some that could be explained by either model. The first column would ask for a summary of the video: "What happened?" The second and third columns were reserved for the competing models. In each of these boxes, students would find prompts: "How well does it match?" and "Evidence?"

After watching five videos and assessing each model's fitness against the evidence, I would have students answer the following: "Based on what you've seen so far, which model do you agree with most as a way to explain sound waves: model 1 or model 2? Explain why you've chosen that model."

Finally, I would ask students to watch one last video of a tuning fork being struck, then dipped in a cup of water (and filmed in slow motion). I used the predict-observe-explain format that I hoped would help students relate their chosen model to another observed sound phenomenon.

It was finally time to enact the lesson. Toward the end of the unit, I asked students to copy the two model descriptions and drawings into their lab notebooks, which they use as textbooks for the course. I left the poster up for the remainder of the unit.

After briefly discussing both models, my collection of *data* began with observations of the students during the lesson, as well as student responses to my tool. I gave students the sound comparison tool and began showing the videos. We filled out the first row (showing a person breaking a wine glass with his voice) as a class to help them get a sense of what each part of the table was asking them to do (see the sample student response in Figure 7.8). Some students commented that lots of energy in a wave could break the glass, whereas others pointed to the slow-motion shots of the glass bending and vibrating. Most were equivocal about which model was best at this stage, but we wrote down our observations and continued on.

The last video before they were asked to choose between models was of a bell in a vacuum chamber as the air was removed. I had deliberately chosen this video as the "clincher"; I wanted to show evidence that sound required a medium, but I didn't want that evidence to be examined until students were already building a conception of which model might work best. I waited for students to point out that, if sound stopped working when the air (medium) was removed, that was strong evidence for model 2.

FIGURE 7.8 Sample student responses

a.

Which Sound Model Is "Sound"?

For each video, in the table under "What happened?" write a few words about what you observed. Then describe how well you think your observation matches each of the models for how sound works and proved evidence for how it does or doesn't match the model.

Video	What happened?	#1: Sound as Energy Wave	#2: Sound as Vibration
Breaking a wine glass	• Glass broke • Vibrating glass, bending • The straw in the glass bounced around	How well does it match? Wine glass or any glass can break Evidence: Energy waves make nodes and antinodes	How well does it match? the wine glass vibrated a lot. Evidence: straw bounced out and the edges of the glass moved.
Color sound	• 4 colors of sand • spreads out over the table • sand makes different shapes	How well does it match? the sand could move Evidence: Sand particles move around, not through	How well does it match? I don't see how sand could vibrate Evidence: Sand just moves, but it doesn't shake.

b.

Vacuum chamber	• the alarm went off • As the alarm went off, it gets quieter	How well does it match? the alarm couldn't go off. Evidence: the air doesn't go through to make the alarm righer.	How well does it match? Alarm goes off from vibration. Evidence: the alarm vibrated.

Based on what you've seen so far, which model do you agree with most as a way to explain sound waves: Model #1 or Model #2? Explain why you've chosen that model.

Based on what I've seen so far, I agree with mostly model #2 because each video shows more vibration than waves.

Based on your model, predict what you think will happen when a tuning fork is struck, then dipped into a cup of water.

Tuning fork in water	Prediction: I predict that the tuning fork will burn up when dipped into a cup of water.	Observation: the water splashed out. The tuning fork moved in slow motion.	Explanation: the vibration of the tuning fork attracted the water to make the water splash.

What I saw and heard convinced me that the students understood the format and expectations of the activity, which made me hopeful that the written work would at least inform me about my goal! Indeed, the written work displayed a variety of depths of student understanding. All the students chose model 2 at the end. In terms of my goal, I was able to *infer* that, given two competing models and evidence to discriminate between the two—and given class discussion to clarify what was observed in each piece of video evidence—students were able to choose the better-supported model. This was quite a lot of scaffolding, but as I mentioned, that was intentional.

Interestingly, my assessment also provided data about students' proficiency with another skill in scientific reasoning: justifying statements with evidence. As it turned out, while the students approached the original goal quite well, they had a great deal of difficulty with choosing and explaining appropriate evidence.

On reflection, I was able to "unpack" some of the skills involved in using evidence. Students have to not only be able to watch, understand, and summarize what happened in the video; they also need to predict what *would be observed* given each model, evaluate their observations against the predictions, and provide specific observations from the video that distinguish between predictions. That's some sophisticated cognition!

This is not to say that I have lowered expectations about what my students can do. Rather, I found firsthand something that I recalled from my science education course work: Large suites of tasks that scientists engage in, such as argumentation from evidence, are not ones that we humans are born knowing how to do. Engaging in Argument From Evidence is one of the eight science and engineering practices identified in the *NGSS*, meaning it can't be mastered in a single lesson, a unit, or even a year; it must be taught and honed, little by little, by running its elements through many years of science course work. In fact, the predict-observe-explain cycle that I asked students to undertake at the end of the activity was a kind of stepping stone and practice for evidence-based argumentation.

What my intuition had told me about teaching students about models—that it was best to start small and simple before handing them the reigns—holds equally true for the use of evidence, a skill that I had built into my assessment without much thought. What I see before me is an endless opportunity for refining my own practice and my students' knowledge and abilities with an infinite series of feedback loops. I summarize my trip through the feedback loop in Figure 7.9 (p. 136).

CHAPTER 7

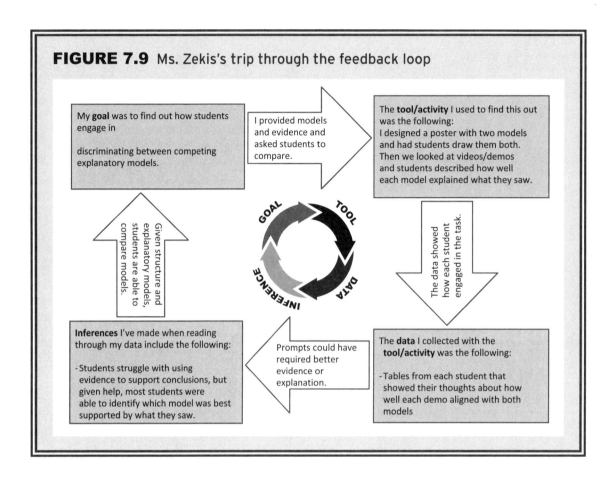

FIGURE 7.9 Ms. Zekis's trip through the feedback loop

Using the Feedback Loop to Plan and Inform Instruction

RESOURCE ACTIVITY 7.1
Reflecting on Previously Collected Data

The purpose of this activity is to guide you to reflect back on an old set of student data using the elements of the Feedback Loop. Start by summarizing the data you collected in the box in the lower right hand corner. Then, take a few moments to read through each piece of data and consider the responses to each of the questions in the boxes below.

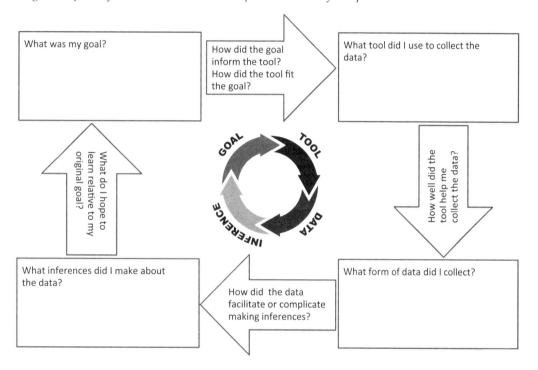

1. What was my goal in collecting these data?

    a. How did the goal inform the tool?

2. What tool did I use to collect the data?

    a. How did the tool fit the goal?

    b. How well did the tool help me collect the data?

THE FEEDBACK LOOP Using Formative Assessment Data for Science Teaching and Learning

CHAPTER 7

    3.   What form of data did I collect?

        a.   How did the data facilitate or complicate making inferences?

    2.   What inferences did I make about these data?

        a.   What did I learn relative to my original goal?

Once you have responded to each of these questions, take a few minutes to consider the take-aways from this experience. Use the following questions to guide your reflection:

- Did I have a goal that was aligned to what I really wanted to know?
- How might I have identified a better tool to assess the goal?
- How might I have limited the amount of data I collected to help me better respond to the goal?
- How did the inferences I originally made relate to the goal?
- What did I do with this information?

Finally, take a few moments to look forward using what you learned in this activity.

What have I learned about
- setting learning goals aligned to what I really want students to know and be able to do?
- selecting streamlined tools that are well aligned to my goals?
- collecting just enough data to inform how well students met my goals?
- making inferences that will guide my next steps for instruction?

Using the Feedback Loop to Plan and Inform Instruction

RESOURCE ACTIVITY 7.2
Planning for Instruction With the Feedback Loop

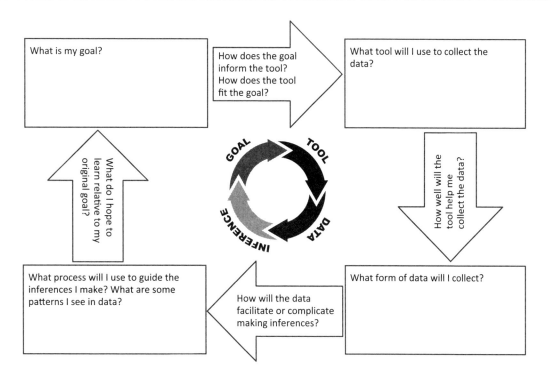

This activity will help you to make a deliberate plan for collecting and analyzing data about teaching and learning in your classroom. Use the questions below as a guide to making a plan.

Overarching idea: What disciplinary core idea, crosscutting concept, or science practice, teaching practice, or other theme is the lesson about?

Goal: What do I want to find out about my teaching of the lesson and/or student learning?

Tool: What tools am I going to use, adapt, or create to generate information relative to my goal?

Data: What data will I collect with the tool I have developed? When and how will I collect it? How much will I collect?

Inferences: What interpretive frameworks are available or can I plan in advance to help me make inferences on the basis of my data?

CHAPTER 7

Reference

Vanides, J., Y. Yin, M. Tomita, and M. A. Ruiz-Primo. 2005. Using concept maps in the science classroom. *Science Scope* 28 (8): 27–31.

CHAPTER 8

Collaborating With Colleagues

It's good to be reminded of how talented, insightful and creative your colleagues are ... [it] has allowed us biology teachers to become a unified group again who share ideas and use each other as sounding boards. This involvement not only helps improve my teaching but also made me remember how great and rewarding it is to be a part of a professional learning community.

—*Experienced high school biology teacher*

Even though we're constantly surrounded by people at school, teaching can still be a solitary activity. Although you might bump into colleagues in the hallway, at lunch, or before or after school, many—or most—of us operate primarily on our own as we plan lessons, write assessments, and make decisions to support student learning. When we do have time to talk with colleagues, it's often spent on new policies or disseminating the decisions of committees. When we finally have a chance to talk about teaching, other topics might take priority, such as discussing the timing of labs so that materials can be shared, making quick decisions about end-of-unit tests, or talking about students. This "egg-crate" problem of teaching, in which every teacher stays more or less in his or her own classroom (Lortie 1975), is part of the very culture of teaching in the United States. Fortunately, things are changing in U.S. schools, and it's becoming more common for schools to build in regular times, such as late-starts, to create opportunities for teachers to meet and plan together.

In this chapter, we will discuss the value and importance of working with other educators, and identify ways to go about developing such relationships and opportunities to collaborate. Additionally, we will talk about the power and influence of making your practice public and

CHAPTER 8

how exploring and sharing your data can benefit your teaching and students' learning. Through these discussions we will share procedures and protocols for engaging in these conversations and how you can gain even greater power from the Feedback Loop when working with others.

Why Collaborate?

Each of us has unique insight into our classrooms. No one else brings our perspective, background, and knowledge to our classes. We know our teaching and students in ways no one else does; however, other educators also employ unique views when looking at a classroom and can reach different understandings of what is transpiring in your room and what your students understand or can do. Our colleagues can provide additional ways of looking at student behaviors and teacher moves. Bringing more perspectives to data and classroom events can give you greater confidence in your assessment of student understanding and next steps.

The research on teacher learning is pretty conclusive: Working with colleagues on a regular basis is key to realizing lasting changes in your teaching practice. It helps you talk about new ways of doing your work in the context of your classroom and school (McLaughlin and Talbert 2006). Teacher learning communities—or groups of teachers from a single content area who meet regularly to talk about teaching and learning—can serve as communities for critical reflection on teaching, as places to discuss and take risks, and as forums to get feedback to support changes in teaching practice (McLaughlin and Talbert 2001; Talbert and McLaughlin 2002).

Although teachers may form and facilitate their own learning communities at their school sites, such groups may also be guided by a facilitator, such as a university researcher (e.g., van Es and Sherin 2009) or instructional coach (Borko et al. 2008; Cobb et al. 2003). Facilitators can help guide interactions among teachers (Whitcomb 2013; Gröschner et al. 2014), engage teachers in active learning strategies (Penuel, Gallagher, and Moorthy 2011), and provide explicit instruction in new instructional approaches that are more effective at improving students' science learning (Penuel et al. 2011; Gröschner et al. 2014).

Finally, collaboration is a big buzzword today—and not only in education. Working well with other people and in groups or teams is highly valued in multiple professions and can be helpful in many personal interactions, too. Specifically, in science, teaming up in lab groups and scientific communities has been an important means of exploring, sharing, and testing ideas. It is also a way in which scientific results and discoveries are vetted and reviewed or assessed by peers. Similarly in education, and with our focus on the Feedback Loop in particular, working with other educators can be an incredibly valuable way to explore teaching and understand more about students' learning. Through incorporating additional perspectives around *goals, tools, data,* and *inferences,* you can generate additional insights and greater confidence in ones that are voiced.

Additionally, through trying to cultivate productive group interactions and relationships among colleagues, we can better relate to what we might often ask our students to do. In many schools and classrooms today, there is an emphasis on group work and giving students experience working with diverse sets of people toward common goals. The group work might be on a lab or project, and there can be varied reasons or wants for supporting these interactions in your classroom. However, although we might encourage our students to work collaboratively and we can talk about how these experiences could benefit them, we might have few opportunities to do that ourselves and gain or explore the benefits of those interactions. Through making efforts to collaborate with colleagues, we can aid our growth as educators and could gain insight into group work in general—whether it's with colleagues or what we ask our students to do.

A powerful result of collaboration is that we make our practice public. It can be easy to work in ways that allow few people outside your room to know what occurs within it. This reality can save us from scrutiny or potential criticism. Additionally, we understand that varied teacher evaluation systems can lead teachers to feel it is risky to open their practice to a wider audience. However, we believe that granting people greater insight into our teaching can benefit our own teaching, others' teaching, and many students' learning. It's through the sharing of our experiences and analyzing the data that arise from our classes that we can develop deeper understandings of how students learn and what practices lead to meeting goals for us and our students.

Although collaboration is powerful, it takes work to develop successful collaborations with other people. The next sections provide more information on ways to cultivate powerful group dynamics that can support your practice.

Getting Started

Whether you're already working at a school that provides time and structures for teachers to work together in learning communities or you're at a school where you work primarily by yourself, we find it useful to explicitly talk through the different steps a group of colleagues might go through to establish a learning community and then to take up the Feedback Loop.

Engaging Your Colleagues

If there are not already pre-established learning communities at your school, you'll have to get one started yourself. This can be as simple as approaching one colleague in your department and asking her or him to look at some of your data together. You may either work with colleagues in the same subject area (e.g., high school biology) or in groups with different levels and types of science represented. With the former, you have the advantage of working with those who will be doing the exact same thing as you, so the activities you develop can be the same. If you work at a small school and are the only one in your

CHAPTER 8

subject area, you might partner with teachers at neighboring schools. The latter approach can involve teachers from across content areas, either at your own school or other schools, even across the country.

We've found that a low-risk way to initiate this relationship is to identify one or two people and ask them to look at some student data with you. A simple statement such as, "I had my students sketch out their initial models for the evaporation of water, and—wow—there are a few I can't make sense of. Would you have time to sit with me and look through them?" Usually, a statement like this appeals to a colleague's expertise in science, as well as their skills as a teacher, and we've heard of very few instances when it's not successful. Once that first session is over, you might casually suggest the colleague work more formally and more regularly alongside you. Invite the colleague to bring data, too; you could say, for example, "If you have anything that you ever would want to look at together, I'd love to chat with you about it."

Getting Time to Meet

Chances are that your time at school is extremely limited, and it is likely already designated for all manner of things that may or may not align with the Feedback Loop; however, we suggest setting a regular meeting time and keeping to it. Sixty to ninety minutes once or twice a month has worked well for teachers in our projects. If you have regular professional development or planning time built into the calendar at your school, you might check with your administrators to see if you can dedicate that time to working with the Feedback Loop.

One school Erin worked with had no common planning time when she first started with them. When the teachers approached the principal for extra time, they were told there was sufficient demand for biology classes that there was no period of the day that all teachers could be free. So, to get started, the biology team decided to set time after school for their learning community meetings. They met regularly for two years and collected data of their students' engagement in new activities and also formalized testimonials of their experiences. As one teacher wrote, "I got involved because I wanted the opportunity to collaborate with colleagues, and I found that planning with other people brings out better ideas, activities, [and] understanding."

After they had established the effectiveness of their learning community time, the teachers went back to their principal and shared their experiences, as well as summaries of some of the data they had collected that showed improvements in student performance. To their surprise and delight, he offered to reshape the schedule for the upcoming academic year to embed planning time in the regular schedule every day for the team.

Setting Norms

Whenever a group comes together, it is important to acknowledge that people have different backgrounds and often express varied beliefs and perspectives. Many of us have varying educational philosophies, and even though we might all share similar goals for our students, rich, powerful collaboration does not automatically arise by putting a group of well-intentioned educators together. Regardless of how well you get along with your colleagues, or how similar or different you are in terms of philosophy, values, beliefs, and background, it can be important to explicitly talk about how and why you would like to collaborate.

Garmston and Wellman's (1999) seven norms of collaboration are a helpful guide for establishing standards for teachers working together with the Feedback Loop; they are summarized in Table 8.1. These guidelines can help develop an environment in which you and your colleagues can safely discuss events in your classroom and ground your feedback and observations in data.

TABLE 8.1 Seven norms of collaboration

NORM	DESCRIPTION
Pausing	Allowing time between when a question is asked or after someone speaks, or allowing time for ideas to settle
Paraphrasing	Reflecting what a speaker has said at the beginning of a new comment to value her or his contribution; for example, "So what you're saying is …"
Putting inquiry at the center	Setting a collective intention to "explore the perceptions, assumptions, and interpretations of others before presenting or advocating one's own ideas" (p. 34)
Probing for specificity	Asking for more information after vague comments are used (e.g., if a teacher says, "They want us to do common assessments," a probe might be, "Who do you mean by 'they'?")
Putting ideas on the table	Throwing ideas and data about student learning out for consideration by the group
Paying attention to self and others	Being aware of what you are saying as well as what others are saying and feeling
Presuming positive intentions	Always assuming that the intentions of others are positive

Source: Adapted from Garmston and Wellman 1999.

We also suggest that you and your colleagues set an additional norm: pick one person who will be responsible for setting an agenda and stick to that agenda. We've all attended meetings that stutter to get started or that meander once they get going, with group members drifting off-task as others try to remember what the project was for the day. When you have

CHAPTER 8

limited time to work together, it's essential to get down to business immediately and not spend time deciding what you'll do on a given day. At the end of every meeting, decide what you'll do next time, who will do what in between meetings, and which person will be responsible for keeping the agenda. You can also refer to Chapter 9 for a number of resources that can get you started on developing norms and additional protocols that can get your group started on the right path.

Engaging in the Feedback Loop With Colleagues: A Four-Meeting Plan*

So far in this book, you have seen examples of the Feedback Loop on a number of different timescales. Mr. Carlson (pp. 47–52) used the Feedback Loop to reflect on and revise an assessment over the course of the month, while Mr. Graham (pp. 81–84) and Dr. Morrison (pp. 95–100) showed us how a teacher can run through multiple cycles in a single day or class period. But the Feedback Loop can also be a guide for teachers over much longer periods of time, such as whole units and years of instruction.

Erin's research has explored how a process like the Feedback Loop can guide how teachers examine and work with representations of student thinking on these much longer timescales. This size of cycle is clearly much slower, but it allows teachers to work collaboratively over a longer period, plan whole units of instruction together, and develop and reflect on multiple formative assessments at the same time.

This adaptation of the Feedback Loop focuses more explicitly on what teachers can do when they are working together to iteratively set learning goals, design tools, and collect data, and to make inferences on the basis of that data to guide their instruction. This process is an adaptation of the Feedback Loop that describes more definitively what teachers can do when they are working together to take up resources such as the *Next Generation Science Standards* (*NGSS*; NGSS Lead States 2013) and use them as a basis to iteratively design, enact, and revise common formative assessment tools. These tools are *common* because the teachers in a learning community design them together after setting goals. Then, each teacher collects data in their own classrooms, and finally, they all come back together to make inferences and identify next steps for their teaching (Ainsworth and Viegut 2009).

In the sections below, we will walk you through a guide to how you and your colleagues can go through the Feedback Loop in the space of four meetings. This cycle can be stretched out into longer periods of time or compressed; however, we suggest that when you and your colleagues are first learning about the Feedback Loop, you take approximately a month to go through the entire cycle. We summarize the steps in Figure 8.1.

This suggested process is informed by the results of two of Erin's funded research projects, which have engaged departments of science teachers from four high schools in this process

---

* This section is based on a paper coauthored by Sara C. Heredia.

FIGURE 8.1 Guide to using the Feedback Loop with colleagues

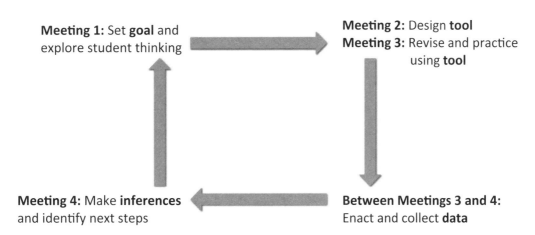

over the course of several school years. In these studies, we have found that engaging in the Feedback Loop leads teachers to develop formative assessments of increasing quality, to synchronize their units, and to learn more about science and student thinking. At the same time, we have observed that student learning increases the more teachers use the Feedback Loop to guide their formative assessment design (e.g., Furtak and Heredia 2014; Furtak, Morrison, and Kroog 2014).

Meeting 1: Set Goal and Explore Student Ideas

At the first meeting, start by setting your *goal*. It may come directly from standards you follow or from other resources available at your school; for example, you may refer to pacing guides or curriculum frameworks adopted by the state, district, or school. Make sure you have all of these resources nearby before the meeting, or gather in a place where you can easily get what you need.

Use the resources provided in this book to help you unpack the disciplinary core ideas, science practices, and crosscutting concepts in the standards; then, step back and map out your own understanding of the idea or practice and how you currently teach it. What activities do you engage students with in support of this idea? What connections do you currently make with crosscutting concepts? Which science practices do students participate in as they learn this idea? Starting with your current methods will help you and your colleagues get on the same page about what you currently know and do.

At the end of the first meeting, you should be able to have a well-developed goal in mind, a listing or representation of common student ideas about it, and a sense for how these concepts relate within a unit.

CHAPTER 8

Meeting 2: Design Tools

Now it's time to find—or design—a common *tool* that will help you to reveal student thinking. Think outside the box, and don't be afraid to throw out a range of different ideas. Often this meeting involves looking for resources in books and online sources such as Trauth-Nare and Buck (2011) or Keeley's formative assessment probes (e.g., Keeley 2008). You could also work with Furtak's (2009) process for designing formative assessment tools. If someone has a computer, live draft a tool as you come up with it. Haydel and colleagues' vignette (pp. 153–157) also describes a good process to follow to look at a number of pre-existing assessment tasks to determine if they would work as good tools aligned with your goals in the Feedback Loop.

Meeting 3: Revise and Practice Using Tools

The next meeting usually begins with the teacher who drafted the tool sharing her or his draft with the group, and the group going through and suggesting revisions. This process can be based not only in your own reading of the tool and how you might respond to it but also on what you think students will say. Go back to Resource Activity 3.2 (p. 66) to guide your conversations about the quality of the tool. In addition, look again at the research you identified into students' common everyday ideas or learning progressions from the *NGSS*, and use them to identify the responses that you might expect to see or hear in response to the assessment you designed. Resource Activity 3.3 (p. 67) can help you formalize the responses you expect students to share and guide a conversation about how you might respond in turn. Revise your assessment if necessary to reflect what you anticipate will happen.

Also take some time to practice using the tool. It can be intimidating to try out a new tool with your students, but if you take some time to envision how the activity will go in your classroom, you're more likely to be successful. When in the class period will you use it? Will students work on it individually or in groups? Will you hold a whole-class discussion of their responses? If so, how will you set it up? What kinds of responses do you think students will share, and how will you take up and work with those responses? We've found it to be important to have frank discussions with colleagues at this point. What about using this new assessment makes you uneasy? What additional approaches or strategies would help you give it a try?

Between Meetings 3 and 4: Enact and Collect Data

Now comes the fun part—every teacher enacts the activity in her or his classroom and collects *data*. If you're working in the same content area, this might happen about the same time. If you're working with colleagues teaching other science courses, it might be more asynchronous. Either way, be sure that you have created a way to capture the student thinking that is shared, such as by collecting written student work, jotting student ideas on a piece of paper, taking a picture of the whiteboard where they have sketched their ideas, or saving the digital file into which their ideas were captured or deposited. Making a video recording of your lesson can also be a wonderful source of data to share with your colleagues to guide your discussions of student ideas and teaching strategies (Furtak, Morrison, and Kroog 2014).

Meeting 4: Make Inferences and Identify Next Steps

Once you have enacted the activity, come back together with your colleagues and dig into the data you collected to make *inferences* together. This can be done formally, as with a protocol such as the one we included in Resource Activity 8.1 (p. 158), or informally. If you have copies of student work, pair off and look through student responses, sorting them into piles when possible, and talk about what students are really thinking and how that relates to the ideas, concepts, and practices you are assessing. You'll be astounded by what you can learn by focusing in great detail on just a handful of responses.

If you're watching a video, since your time together is short, pick a brief clip (two to five minutes) in which an interesting student idea was surfaced and watch it together. Talk about what is happening in the video, your own understanding of it, and how this instance applies to broader principles of teaching and learning (e.g., Sherin and van Es 2003). Reflecting on your own video can help you focus more intently on student thinking (Sherin and van Es 2008).

Think back to step 2, when you anticipated student responses, and compare those expectations with the answers students actually provided. Are there discrepancies between what you expected and what students actually said or wrote? What surprised you? Avoid the common trap of noticing what students did *not* provide and instead focus on the information students *did* give you about their thinking.

As you make inferences, hold each other accountable for deciding *what happens next*; that is, given the patterns you are seeing in the data, what will you do in class to help students advance in their learning? As you interpret student responses, you will also glean insights about what to do next time. Write these down somewhere you will remember. If you want to revise the activity, do it digitally now, or at least capture handwritten edits for next time. Then, think about your next trip through the loop: What new goals emerged for you? Did the process suggest developing a formative assessment for a related standard or for something else entirely?

We suggest using either Resource Activity 5.1, "Guide to Tracking Inferences" (p. 101), or Resource Activity 6.1, "Multiple Feedback Loops" (p. 119), as a way to structure your processes of making inferences and identifying next steps. What to do next often emerges organically while engaging in these processes. As you come up with ideas for new goals, be sure to jot them down to bring to the next meeting.

CHAPTER 8

Collaborative Tool Design in a Teacher Learning Community

Erin Marie Furtak and Sara C. Heredia, University of Colorado Boulder

The biology teachers at Sagebrush* High School were concerned that students might not be focusing on the major takeaways of this multiple-day activity, which was aligned with *NGSS* HS-LS4-3 and HS-LS4-4. This activity, familiar to many biology teachers, involved students using different types of spoons—wooden, slotted, and those used as plastic ice cream samplers—to represent variations in beak shape and size in Darwin's finches on the Galápagos Islands. Students used these spoons to collect mixtures of seeds from "wet" and "dry" years, following Peter Grant's research (Grant 1986), and created graphs of the number of individuals with each beak type at the end of multiple wet and dry years.

In their first meeting, the teachers identified a *goal* of determining if students were able to distinguish patterns and cause-and-effect relationships between variation in a single trait of one species and the change in proportion of that trait within the population as environmental conditions altered.

At the second meeting, the teachers adapted a set of discussion questions they usually assigned students after the activity into a *tool* that would help draw out student ideas about the activity. This tool was designed to draw out students' conclusions on the basis of engaging in the activity as well as their everyday explanations about how organisms are able to change themselves in response to environmental alterations. Once the questions were in draft form, one of the teachers volunteered to type them up for the next meeting.

At the third meeting, teachers read through the questions carefully, anticipated how students might respond, and made edits to improve the wording. They also had a discussion about the different types of ideas the students might share and how they as teachers might respond to them with activities to move the students forward in their learning. These "feedback activities" helped the teachers be more concrete about how their unit was sequenced to address students' anticipated everyday conceptions as surfaced with the tool.

All of the teachers then engaged their students with the tool. They gave students time to work on the questions individually and then held a whole-class discussion to reveal thinking. Teachers then collected written responses, yielding two forms of *data*: Students' written responses, as well as their comments shared in the whole-class discussion. Students' written responses are summarized in Table 8.2.

TABLE 8.2 Sample student responses to a natural selection tool

QUESTIONS FROM THE TOOL	STUDENT	DATA: SAMPLE STUDENT RESPONSE
1. Remember, all of these birds are from the same species! Based on that knowledge, which population of birds (beak types) were the most successful? Why (i.e., what environmental conditions allowed for this success)?	1	The "plastic sampler" beaked birds were the most successful because the birds had a better beak to carry more food back and forth with.
	2	During both the wet and dry years, the birds represented by the plastic sampler spoons were most successful given the number of birds back at the nest. Though the wet years produced more seeds, this type of bird was capable of surviving in both conditions, with the wooden spoon coming in a close second.
	3	Plastic sampler, the beak, size of food, number of birds.
	4	The plastic sampler was the best because they had the highest number of birds in the nest, and they had the most seeds. They were small and rounded.
2. What is natural selection, and how is it simulated in this activity? Specify something about beak type in your answer. What are these differences among the same species called?	1	Natural selection is "only the strong survive" and is simulated in this activity by the birds with the best "beaks" live.
	2	Natural selection occurs when various conditions are naturally presented to a certain species. Only organisms within that species that have a quality that allows them to survive under those conditions lives to reproduce. Difference within the same species is called "variation."
	3	The random selection of a species. Everyone has a random beak.
	4	Natural selection is when nature selects something to help an animal survive.

The teachers had all made on-the-fly inferences that guided their instruction when they were leading the whole-class discussions, but they also wanted to come together around a set of student work. Thus, at their next meeting, they gathered together to read through a sample of student responses (see Table 8.2) and make some systematic *inferences* about what the students were saying. Rather than work with hundreds of student responses from across all of the classes, they all looked at copies of a subset of students' responses. They made a number of inferences about what the students knew, including that students were able to correctly identify that the birds with beaks represented by the plastic sampler spoon were the most successful, suggesting that they were able to interpret the graphs they had made of the data generated by the

CHAPTER 8

> model. However, they also noted that only some of this subset of students (e.g., students 2 and 4) cited evidence in their responses. At the same time, several students, such as student 1, used language that suggested they were using the word "fitness" in its everyday sense ("only the strong survive"). They also noted that student 3 used the term "random" in unclear ways.
>
> The teachers drew on this subset of student responses to inform their teaching. After jotting down some edits to the tool that they would use next year, they also noted that they wanted to be more careful about identifying the scientific meaning of fitness with students in subsequent lessons. They also talked about what different organisms they could highlight in their class that were not necessarily the strongest but that were more successful.

Multiple Timescales for Collaborative Work: School-Based Teacher Communities

The examples we've seen so far have involved groups of teachers working over the course of several weeks; however, in several settings, we've used the Feedback Loop with collaborative groups on much shorter timescales. For example, teachers might plan an activity using a blank version of the loop in a single sitting (steps 1–3) and then come together for a final meeting after collecting data. This procedure works best once teachers are already familiar with all the steps and have already taught the unit they're working in.

Exploring the *NGSS* Through Task Analysis

Angela Haydel DeBarger, George Lucas Educational Foundation

Christopher Harris, SRI International

William Penuel, University of Colorado at Boulder

Katie Van Horne, University of Colorado at Boulder

A big challenge facing science educators who are shifting their instruction to meet the vision of *A Framework for K–12 Science Education* (NRC 2012) and the *NGSS* is how to assess students' progress toward achieving the new standards. Chapter 2 described this new, integrated vision of what it means to become proficient in science that emphasizes using and applying knowledge in the context of disciplinary practices. This knowledge-in-use perspective holds that disciplinary core ideas, science and engineering practices, and crosscutting concepts together enable learners to make sense of phenomena or design solutions to problems. Consequently, each of the *NGSS* performance expectations—that is, the actual statement of student learning goals—integrates these three dimensions together.

The new science standards mean that developing tools aligned with goals focused only on acquisition of facts is no longer sufficient. Instead, new tools will be needed to assess integrated science learning. As assessment designers and researchers, we see the *NGSS* as an exciting opportunity to change science assessments so that they really help teachers understand their students' developing proficiencies in science. There is a real need to do this assessment design work well because assessment will play a central role in supporting implementation of the vision set forth in the *Framework* and the *NGSS*.

Unfortunately, there are very few examples of tools that integrate core ideas, crosscutting concepts, and science practices in the manner intended by the *Framework* and the *NGSS*. Science assessment designers are developing new ways to design these assessments (e.g., DeBarger et al. 2014; DeBarger et al. 2015), and this book presents the Feedback Loop as an approach for teachers to use in designing assessments themselves. Our ongoing collaborative research and development efforts with teachers have focused on new models for assessing

CHAPTER 8

next generation science learning that require students to demonstrate the *NGSS* performance expectations and provide teachers with usable and actionable information about students' progress toward achieving these new learning goals.

In March 2015, we led a full-day National Science Teachers Association (NSTA) Professional Learning Institute (PLI) with over 90 science teachers and district leaders on our approach for designing next generation science assessments. The PLI goals were to analyze existing tasks in relation to the *NGSS*, identify the core aspects of science practices and evidence for students to perform those practices, develop an assessment argument to guide design and refinement of tasks that integrate the three dimensions, and discuss strategies for using tasks formatively and intentionally in classrooms to elicit and build on students' integrated learning. You can view all the materials from the PLI at *http://learndbir.org/talks-and-papers/nsta-2015-pdi-developing-next-generation-science-assessments*.

Task analysis was the first activity of the day, following a brief introduction to the *Framework* and the *NGSS*. This activity focused on a single performance expectation: MS-LS4-4: Construct an explanation based on evidence that describes how genetic variations of traits in a population increase some individuals' probability of surviving and reproducing in a specific environment (see Figure 8.2). Prior to the PLI, we identified six tasks related to the disciplinary core idea of natural selection represented in this performance expectation. We found the tasks on various websites, including some state departments of education, teacher-created websites, research organizations, and companies that develop curriculum materials. We wanted to provide a few contrasting cases that would provoke conversations about how well the performance expectation was met. We included tasks with the following characteristics: high cognitive demand without addressing the science practice, extended inquiry with limited integration of the core idea, conceptually oriented multiple-choice with distractors that focus on reasoning, and a well-blended example.

Collaborating With Colleagues

FIGURE 8.2 *NGSS* performance expectation MS-LS4-4

MS-LS4 Biological Evolution: Unity and Diversity		
MS-LS4-4.	**Construct an explanation based on evidence that describes how genetic variations of traits in a population increase some individuals' probability of surviving and reproducing in a specific environment.** [Clarification Statement: Emphasis is on using simple probability statements and proportional reasoning to construct explanations.]	
Science and Engineering Practices	**Disciplinary Core Ideas**	**Crosscutting Concepts**
Constructing Explanations and Designing Solutions Constructing explanations and designing solutions in 6–8 builds on K–5 experiences and progresses to include constructing explanations and designing solutions supported by multiple sources of evidence consistent with scientific ideas, principles, and theories. • Construct an explanation that includes qualitative or quantitative relationships between variables that describe phenomena. (MS-LS4-4)	**LS4.B: Natural Selection** Natural selection leads to the predominance of certain traits in a population, and the suppression of others. (MS-LS4-4)	**Cause and Effect** Phenomena may have more than one cause, and some cause and effect relationships in systems can only be described using probability. (MS-LS4-4),(MS-LS4-5),(MS-LS4-6)

This task analysis activity mirrors the work that many teachers will have to do to implement the *NGSS*, as the tasks we selected for this activity were created prior to the new standards and would need to be revised to address the *NGSS*. While developing a comprehensive *NGSS* assessment will need to address a "bundle" of performance expectations, we focused on one for this activity. In doing so, our intention was to facilitate teachers' learning about the integration of a disciplinary core idea with a practice and crosscutting concept. We also wanted to find out what criteria teachers used to evaluate the quality of different assessment tasks relative to an *NGSS* performance expectation. Our aim was to provide a grounded means for engagement with the dimensions of the performance expectation.

Working in small groups, participants reviewed the assessment tasks, rated them with respect to how well they assess the three dimensions integrated in the performance expectation, discussed their ratings, and recorded their ratings in an online form. For each task, participants were asked two questions: (1) How effective do you think the task would be for eliciting three-dimensional science learning? (five-point scale from 1 = not at all effective to 5 = highly effective), and (2) What are the reasons for your rating? This is an activity that teachers could engage in at their school sites very easily.

Because the teachers' responses were collected online, we could immediately share the results during the PLI, and the ratings provided excellent grist for a whole-group discussion. In reviewing the ratings together, we prompted

CHAPTER 8

participants to identify some general criteria for what constitutes a three-dimensional task and also consider what the ratings reveal with regard to differences in how we are understanding the dimensions differently. Generally, tasks that focused primarily on the disciplinary core idea were recognized by most groups as inadequate for meeting the performance expectation. Groups were split in their ratings of tasks that addressed aspects of the science practice but went beyond the performance expectation in terms of understandings related to the disciplinary core idea. One task that asked students to select a response to explain their reasoning also received mixed ratings and prompted a discussion about whether constructed response should be required when a performance expectation calls for an "explanation." While the well-blended task example that we provided fared well (high consensus that it met the *NGSS* performance expectation), it raised other concerns about language and reading demands that may interfere with students' ability to demonstrate their understanding.

Beginning with this task analysis activity was beneficial for our PLI in a number of ways. First, it provided a lively way to engage with the *NGSS*. The activity required participants to wrestle with issues such as what content is covered in the disciplinary core idea at a particular grade level and what constitutes a scientific explanation. The activity also helped participants really "see" the integration of disciplinary core ideas, practices, and crosscutting concepts as represented in a performance expectation. The idea of three-dimensional assessment can remain obscure and abstract without examples that illustrate, if even only partially, what this integration might look like in an assessment task. The activity also sparked participants' interest in design principles for generating more tasks like those that they were intuitively able to recognize as more aligned with the *NGSS*. Thus, the task analysis helped lay the groundwork for engagement in the assessment design process that was the focus of the rest of the PLI.

This kind of task analysis corresponds well with the goal–tool alignment phase, as described in this book. Looking at a number of assessment tasks that you might use as tools to meet *NGSS*-aligned goals can help you and your colleagues better understand the demands of *NGSS* assessments, as well as develop a shared understanding of how they might be systematically elicited by tools. Reviewing alternative tools that differ in the degree to which they map to an intended goal sheds light on the different ways that tools can elicit a core idea, practice, and crosscutting concept and the qualities of tasks that not only elicit these dimensions but require their integration. In general, constructed response tasks fared more favorably in this regard in our activity because the science practices demand sensemaking. However, we also acknowledged the value of using multiple-choice questions to elicit reasoning and prompt evidence-based arguments as part of class discussions.

> Reviewing a range of tools is also helpful for clarifying what evidence to look for in student responses that would fully address a learning goal (e.g., the components of what constitutes a complete explanation or the level of detail needed to demonstrate understanding of the disciplinary core idea). Working collaboratively in design teams, teachers can document these decisions and use them to guide more consistent tool refinement and generation.

The Feedback Loop in Preservice Teacher Education

This chapter has focused on inservice teachers working collaboratively in the Feedback Loop, but the Feedback Loop can also be a useful tool for preservice teachers to organize and reflect on their own teaching with the support of their mentor teachers, instructors, and peers.

In Erin's preservice science teacher education course, students are asked to plan, enact, and reflect on two lessons in cooperation with their mentor teacher during the course of the semester. The Feedback Loop has served as an orienting framework for this process: It gives teacher candidates an explicit mechanism for discussing a goal with their mentor teacher around which they will develop a lesson and then makes clear the processes of tool selection or design and the planning process for enactment. As the candidates move closer to teaching their lesson, they share their loop with Erin and peers in class to get feedback on goal–tool alignment and their plans for collecting data in their classrooms.

When it's time for the candidates to give their lessons, they collect multiple sources of data of their enactment, including, when possible, videotapes of whole-class discussions and feedback notes from their mentor teachers, university supervisor, or peers. They bring these sources of data back to their college course and use a guided protocol, such as the one included in Resource Activity 8.1, to engage in a structured debrief with a small group of students. This protocol guides the candidates in looking carefully at the data together, and making warranted inferences about how the lesson went and what the students learned. At the same time, the candidates collectively identify next steps to move students forward in their learning, as well as goals for the teacher candidates' professional growth.

Summary

This chapter has provided a rationale, as well as a number of processes and procedures, for engaging your colleagues in the Feedback Loop. We've found these processes to be useful in multiple settings, including school-based learning communities, monthly district professional development meetings, focused cross-district meetings, and teacher education. The bottom line is that working with colleagues not only makes this process more enlightening but more effective and enjoyable.

CHAPTER 8

RESOURCE ACTIVITY 8.1
Collaborative Feedback Loop Protocol

This resource activity is one example of a protocol that can be used with your colleagues to share and discuss the data you might collect with your feedback loop and the inferences and next steps you might take as a teacher.

1. The presenting teacher describes the class, students, and unit they are working in/with. Then she or he shares her or his *goal* with the other teachers in the group. (2–3 min.)

2. The presenting teacher briefly describes the *tool* and how it was designed to align with the goal, as well as the types of responses the teacher expected students would share. (2–3 min.)

3. Participating teachers ask clarifying questions about the *goal* and *tool*. (Group members ask questions that have factual answers to clarify their understanding of the data, such as, "For how long did you collect the data?" and "How many students did you work with?") (1 min.)

4. The presenting teacher shares her or his *data* with the other teachers. Everyone takes a few minutes to read through the data, sharing things that they notice about it (see possible prompts below). This is NOT the time to make inferences or conclusions but just to surface things they notice. The presenter can answer clarifying questions but is otherwise quiet and taking notes. (5–10 min.)

   - A pattern in the data I noticed was _____.
   - While some of the students answered _____, a few other students answered _____.
   - I was surprised to see _____.

5. The participating teachers now make *inferences* about the data, relating it back to the goal. When relevant, teachers link these inferences to next steps they might suggest for the students on the basis of the inferences they are making. (5–7 min.)

6. The presenting teacher summarizes what she or he learned from the process, with an emphasis on how their experience might inspire *next steps* for their teaching (e.g., what she or he will do next on the basis of these data, both in terms of feedback for the students and how she or he might revise the goal or tool for next time) across multiple timeframes, including those listed below. (5 min.)

   - If I could have five more seconds with these students?
   - If I could have five more minutes with these students?
   - When I meet with the students next class?
   - When teaching these students next week?
   - When I move into the next unit this year?
   - When I teach this lesson again next year?

7. The group reflects on the process, sharing thoughts about how the discussion worked for the group. (2 min.)

8. Repeat the process for each member of the group.

Individual reflection: After all members of the group have shared their data, reflect on the experience. Record any insights or benefits you found from having other people discuss your data (as opposed to just looking at it alone). When do you think it can be most valuable to explore data with other people? Why? And when and why would you choose to look at data alone?

CHAPTER 8

References

Ainsworth, L., and D. Viegut. 2006. *Common formative assessment: How to connect standards-based instruction and assessment*. Thousand Oaks, CA: Corwin Press.

Borko, H., J. Jacobs, E. Eiteljorg, and M. Pittman. 2008. Video as a tool for fostering productive discussions in mathematics professional development. *Teaching and Teacher Education* 24 (2): 417–436.

Cobb, P., J. Confrey, A. DiSessa, R. Lehrer, and L. Schauble. 2003. Design experiments in educational research. *Educational Researcher* 32 (1): 9–13.

DeBarger, A. H., C. J. Harris, C. D'Angelo, J. Krajcik, C. Dahsah, J. Lee, and Y. Beauvineau. 2014. Constructing assessment items that blend core ideas and science practices. In *Learning and becoming in practice: The international conference of the learning sciences (ICLS) 2014*, vol. 3, ed. J. L. Polman, E. A. Kyza, D. K. O'Neill, I. Tabak, W. R. Penuel, A. S. Jurow, K. O'Connor, T. Lee, and L. D'Amico, 1703. Boulder, CO: International Society of the Learning Sciences.

DeBarger, A. H., W. R. Penuel, C. J. Harris, and C. A. Kennedy. Forthcoming. Building an assessment argument to design and use next generation science assessments in efficacy studies of curriculum interventions. *American Journal of Evaluation*.

Furtak, E. M. 2009. *Formative assessment for secondary science teachers*. Thousand Oaks, CA: Corwin Press.

Furtak, E. M., and S. C. Heredia. 2014. Exploring the influence of learning progressions in two teacher communities. *Journal of Research in Science Teaching* 51 (8): 982–1020.

Furtak, E. M., D. L. Morrison, and H. Kroog. 2014. Investigating the link between learning progressions and classroom assessment. *Science Education* 98: 640–673.

Garmston, R. J., and B. M. Wellman. 1999. *The adaptive school: A sourcebook for developing collaborative groups*. Norwood, MA: Christopher-Gordon.

Grant, P. 1986. *Ecology and evolution of Darwin's finches*. Princeton, NJ: Princeton University Press.

Gröschner, A., T. Seidel, K. Kiemer, and A.-K. Pehmer. 2014. Through the lens of teacher professional development components: The "Dialogic Video Cycle" as an innovative program to foster classroom dialogue. *Professional Development in Education* (September): 1–28.

Keeley, P. 2008. *Science formative assessment: 75 practical strategies for linking assessment, instruction, and learning*. Thousand Oaks, CA: Corwin.

Lortie, D. C. 1975. *Schoolteacher: A sociological study*. Chicago: University of Chicago Press.

McLaughlin, M. W., and J. E. Talbert. 2001. *Professional communities and the work of high school teaching*. Chicago: University of Chicago Press.

McLaughlin, M. W., and J. E. Talbert. 2006. *Building school-based teacher learning communities: Professional strategies to improve student achievement*. New York: Teachers College Press.

National Research Council (NRC). 2012. *A framework for K–12 science education: Practices, crosscutting concepts, and core ideas*. Washington, DC: National Academies Press.

NGSS Lead States. 2013. *Next generation science standards: For states, by states*. Washington, DC: National Academies Press. www.nextgenscience.org/next-generation-science-standards.

Penuel, W. R., L. P. Gallagher, and S. Moorthy. 2011. Preparing teachers to design sequences of instruction in Earth systems science: A comparison of three professional development programs. *American Educational Research Journal* 48 (4): 996–1025.

Sherin, M., and E. van Es. 2003. A new lens on teaching: Learning to notice. *Mathematics Teaching in the Middle School* 9 (2): 92–95.

Sherin, M. G., and van Es, E. A. 2008. Effects of video club participation on teachers' professional vision. *Journal of Teacher Education* 60 (1): 20–37.

Talbert, J. E., and M. W. McLaughlin. 2002. Professional communities and the artisan model of teaching. *Teachers and Teaching: Theory and Practice* 8 (3): 325–343.

Trauth-Nare, A., and Buck, G. 2011. Assessment for learning. *The Science Teacher* 78 (1): 34–39.

van Es, E. A., and M. G. Sherin. 2009. The influence of video clubs on teachers' thinking and practice. *Journal of Mathematics Teacher Education* 13 (2): 155–176.

Whitcomb, J. A. 2013. Learning and pedagogy in initial teacher preparation. In *Handbook of psychology*, 2nd ed., ed. I. B. Weiner, 441–463. New York: John Wiley and Sons.

CHAPTER 9

Resources

Summary of Resource Activities

Each chapter in this book has included different resource activities intended to guide you through your own process of using the Feedback Loop, arranged according to four elements. Once you have learned about the entire Feedback Loop, we suggest that you begin by using Resource Activity 7.2 (p. 139), which guides you through the process of planning for instruction using this method. This is the version of the Feedback Loop that many of the teachers who provided vignettes for this book were using. Then, as you work through each of the elements, you can use the additional resource activities in Chapters 2–6 to structure each step of the process, starting with Resource Activity 7.2 for an overview, and then using Resource Activity 2.1 (p. 34) and Resource Activity 2.2 (p. 35) when identifying a goal; Resource Activities 3.1 (p. 65), 3.2 (p. 66), and 3.3 (p. 67) when developing, revising, or selection tools; Resource Activity 4.1 (p. 85) when planning how to collect data; Resource Activity 5.1 (p. 101) when making inferences; and Resource Activities 6.1 (p. 119), 7.1 (p. 137), and 8.1 (p. 158) when reflecting on those data and determining next steps for instruction. This pathway is shown in Figure 9.1 (p. 164).

CHAPTER 9

FIGURE 9.1 Guide to using resource activities

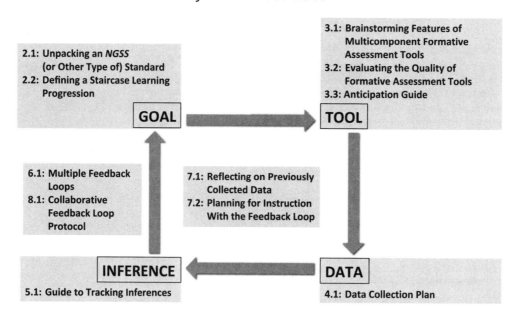

You may feel comfortable using only Resource Activity 7.2 to guide your planning process, and that's just fine; we've found that once teachers learn about the process, the guiding questions in that activity are sufficient. However, if you are using the Feedback Loop with your colleagues, other activities can serve as good starting points for conversations about goals; the assumptions made in tool selection, adaptation, and design; strategies for collecting data in classrooms; and guiding processes of making inferences about data and planning next steps for instruction.

Suggestions for Further Reading

We drew on a number of sources in writing this book, and many of those provide greater detail than we were able to summarize here. We organized the following suggestions for further reading according to elements of the Feedback Loop, as well as collaborating with colleagues and involving students. Although we put these readings into categories, many of them are helpful for multiple steps in the Loop.

Goals

Brunsell, E., D. M. Kneser, and K. Niemi. 2014. *Introducing teachers and administrators to the NGSS: A professional development facilitator's guide.* Arlington, VA: NSTA Press.

National Research Council (NRC). 2012. *A framework for K–12 science education: Practices, crosscutting concepts, and core ideas.* Washington, DC: National Academies Press.

NGSS Lead States. 2013. *Next Generation Science Standards: For states, by states.* Washington, DC: National Academies Press. www.nextgenscience.org/next-generation-science-standards.

Resources

Pratt, H. 2013. *The NSTA reader's guide to the Next Generation Science Standards.* Arlington, VA: NSTA Press.

UCAR. 2015. Science literacy maps. *http://strandmaps.dls.ucar.edu/*

Tools

Ambitious Science Teaching (AST). 2014. *Planning for engagement with important science ideas.* Seattle, WA: University of Washington Department of Education. *http://ambitiousscienceteaching.org/wp-content/uploads/2014/08/Primer-Plannning-for-Engagement.pdf*

Furtak, E. M. 2009. *Formative assessment for secondary science teachers.* Thousand Oaks, CA: Corwin Press.

National Research Council (NRC). 2014. *Developing assessments for the Next Generation Science Standards.* Washington, DC: National Academies Press.

Keeley, P. 2008. *Science formative assessment: 75 Practical strategies for linking assessment, instruction, and learning.* Thousand Oaks, CA: Corwin Press.

Keeley, P., F. Eberle, and C. Dorsey. 2008. *Uncovering student ideas in science, volume 3: Another 25 formative assessment probes.* Arlington, VA: NSTA Press.

Keeley, P., F. Eberle, and L. Farrin. 2005. *Uncovering student ideas in science, volume 1: 25 formative assessment probes.* Arlington, VA: NSTA Press.

Keeley, P., F. Eberle, and J. Tugel. 2007. *Uncovering student ideas in science, volume 2: 25 more formative assessment probes.* Arlington, VA: NSTA Press.

Keeley, P., and J. Tugel. 2009. *Uncovering student ideas in science, volume 4: 25 new formative assessment probes.* Arlington, VA: NSTA Press.

Data

Michaels, S., M. C. O'Connor, and M. W. Hall. 2013. *Accountable talk sourcebook: For classroom conversation that works.* Pittsburgh, PA: University of Pittsburgh Institute for Learning.

Michaels, S., A. W. Shouse, and H. A. Schweingruber. 2007. *Ready, Set, SCIENCE! Putting research to work in K–8 science classrooms.* Washington, DC: National Academies Press.

Inferences

Cartier, J. L., M. S. Smith, M. K. Stein, and D. K. Ross. 2013. *5 practices for orchestrating task-based discussions in science.* Arlington, VA: NSTA Press.

Pellegrino, J. W., N. Chudowsky, and R. Glaser. 2001. *Knowing what students know: The science and design of educational assessment.* Washington, DC: National Academies Press.

Involving Students in Assessment

Coffey, J. 2003. Involving students in assessment. In *Everyday assessment in the science classroom*, ed. J. M. Atkin and J. E. Coffey, 75–87. Arlington, VA: NSTA Press.

Tweed, A. 2009. *Designing effective science instruction: What works in science classrooms.* Arlington, VA: NSTA Press.

CHAPTER 9

Engaging Colleagues

Ainsworth, L., and D. Viegut. 2006. *Common formative assessment: How to connect standards-based instruction and assessment.* Thousand Oaks, CA: Corwin Press.

Blythe, T., and D. Allen. 1999. *Looking together at student work.* New York: Teachers' College Press.

Dana, N. F., D. Yendol-Hoppey, and G. Thompson-Grove. 2009. *The reflective educator's guide to classroom research: Learning to teach and teaching to learn through classroom inquiry.* Thousand Oaks, CA: Corwin.

Dana, N. F., and D. Yendol-Hoppey. 2008. *The reflective educator's guide to professional development: Coaching inquiry-oriented learning communities.* Thousand Oaks, CA: Corwin.

Garmston, R. J., and B. M. Wellman. 1999. *The adaptive school: A sourcebook for developing collaborative groups.* Norwood, MA: Christopher-Gordon.

McLaughlin, M. W., and J. E. Talbert. 2006. *Building school-based teacher learning communities: Professional strategies to improve student achievement.* New York: Teachers' College Press.

National School Reform Faculty. *http://www.nsrfharmony.org*.

Glossary of Key Terms

Alignment: Coherence between elements of the Feedback Loop; for example, goals and tools or data and goals.

Crosscutting concepts: Ideas and concepts, such as Patterns, Cause and Effect, and Structure and Function, that span and unify the disciplines of science and engineering; a component of the *Next Generation Science Standards*.

Data: Information about what students know and are able to do; data can be qualitative or quantitative, as well as formal or informal.

Disciplinary core idea: Concepts and ideas that have two or more of the following characteristics: have broad applicability across a range of sciences, provide a key tool to understanding problems, relate to students' life experiences, and are teachable at multiple levels of complexity over grade spans; a component of the *Next Generation Science Standards*.

Feedback: The process of providing information to improve performance; in the Feedback Loop, the action of connecting inferences and goals to "close the loop" and provide information to move students forward in their learning.

Goals: What you want students to know and be able to do.

Inferences: Claims that identify trends and patterns in data about what students know and are able to do.

Learning community: A form of collaborative teacher working group with the purpose of exploring data to improve the quality of teaching and learning; sometimes called teacher learning communities (in single-content areas) or professional learning communities (often used to describe cross-content groupings, or those involving administrators).

Learning progression: Sequential representations of ideas or practices, increasing in sophistication toward ultimate learning goals for students.

***Next Generation Science Standards* (NGSS):** The newest set of science education standards, developed by Achieve, Inc. and based on *A Framework for K–12 Science Education;* they consist of three-dimensional standards that intertwine science practices, disciplinary core ideas, and crosscutting concepts.

Norms: Shared meanings for acceptable behavior or interaction within a group of teachers or among a teacher and her or his students.

Performance expectation: Descriptions in the *Next Generation Science Standards* that explain what students must be able to do to show they have met a standard.

GLOSSARY OF KEY TERMS

Scientific inquiry: The process of teaching learning by engaging in the activities of scientists.

Science practice: The actions of scientists and engineers as they go about their daily work, including Developing and Using Models, Constructing Explanations, and Engaging in Argument From Evidence; a component of the *Next Generation Science Standards*.

Student ideas as resources: View that students' everyday ideas are important resources to build on for instruction; contrasted with viewing student ideas as wrong or misconceptions that must be replaced.

Tools: How you'll find out what students know and are able to do; the activities that we use to surface student thinking in the Feedback Loop.

Index

Page numbers printed in **boldface** type refer to figures or tables.

A

A Framework for K–12 Science Education, 153, 154, 167
Accountability, 3, 70, 108, 149
Accountable Talk Sourcebook, 45
Alignment, 167
    of tools with goals, 7, 39, 124–125, 153
        for astronomy unit, 47–52, **48, 49, 51, 52**
Anchoring events, 39–40, 46, 56, 66
Anticipating in the Five Practices model, 56
Anticipating student use and responses to tools, 42, 56
Anticipation Guide, 67
Argumentation, 6, 95, 135
Asking questions and defining problems, designing tools for assessment of, 53–54
Assessment(s), 4–5
    common elements of, 6
    development of, 5
    formative, 3, 4–5, 3
        definition of, 4–5
        distinction from summative assessment, 5
        Feedback Loop and, 104–105
        setting norms for conversations for, 57–59
        three-step process of, 104
        tools for, 37–67
    NGSS and, 153–154
    standardized tests, xiii, 3, 7, 69, 70
    summative, 5, 9, 39, 84
Assessment conversations, 44–45
    feedback in, 108–112
    setting norms for, 57–59
Astronomy unit
    goal–tool alignment in, 47–52, **48, 49, 51, 52**
    learning progression in, 26–27, **28–29**

B

Bennett, R. E., 88, 94, 107
Berkeley Education Assessment Research (BEAR) assessment system, 5
Big ideas of science, 19, 30, 39, 66
Bird beak natural selection unit, 150–152, **151**
Biston betularia adaptations unit, 108–112
Black, P., 88
Briggs, D. C., 43
Buck, G., 148

C

Cellular transport unit, 73–76, **74–76**
Clicker questions, **32**, 41, 44, 45, 78, **83**
Climate change, 61–64, **62, 63**
Closing the Feedback Loop, 103–119
    evaluative vs. informational feedback, 105–107
    feedback strategies, 107–113
    formative assessment and the Feedback Loop, 104–105
    identifying the gap between goals and student learning, 103–104, 106
    Resource Activity 6.1: Multiple Feedback Loops, 119
    for sinking and floating unit, **114**, 114–118, **118**
Coffey, J. E., 108
Cohen, E. G., 45
Collaboration with colleagues, 130, 141–159. *See also* Learning communities
    engaging colleagues for, 143–144
    exploring *NGSS* through task analysis, 153–157, **155**
    Feedback Loop in preservice teacher education, 157
    four-meeting plan for engaging in Feedback Loop, 146–149, **147**
        meeting 1: set goal and explore student ideas, 147
        meeting 2: design tools, 148
        meeting 3: revise and practice using tools, 148
        between meeting 3 and meeting 4: enact and collect data, 148–149
        meeting 4: make inferences and identify next steps, 149
    getting started with, 143–144
    meeting agenda for, 145–146
    meeting time for, 144
    Resource Activity 8.1: Collaborative Feedback Loop Protocol, 158–159
    setting norms for, **145**, 145–146
    tool design in a teacher learning community, 150–152, **151**
    value and importance of, 142–143
Concept maps, 46
    about natural selection, 126–129, **128**, 129
Constructed response question, 43, 64, 94, **118**, 156
Content standards in *NSES,* 19–20
Crosscutting concepts, 6, 20, 139, 147, 153
    definition of, 167
    for energy cycling and photosynthesis, **22, 34**, 46

INDEX

for plate tectonics, **96**
for structure and properties of matter, **21**
Cuing, as feedback strategy, 108, 111, 113

D
Data/data collection in Feedback Loop, 3–4, 7, **8,** 69–85, 139
   for *Biston betularia* adaptations unit, 109–111
   collaborating with colleagues for, 148–149
   connections with other elements, 10–12, **11,** 78–79, **79**
   data analysis, 92
   data reduction, 78–79
   definition of data, 7
   disconfirming data, 92–93
   formal data, 3, 7, **77,** 77–78, 79, **85,** 90, 127, 167
   goals of, 4, 78, 80
   going back to, 92–93
   grounding inferences in, 89
   for heat vs. temperature unit, 82
   informal data, 3, 56, **77,** 77–78, 79, 80, 84, **85,** 90, 127, 167
   interpretation and inferences from, 4, 5, 7, 11, 87–101, 125 (*See also* Inferences in Feedback Loop)
   involving students in, 80
   for natural selection unit, 126, 150–151
   planning for data collection, 85, **126**
   for plate tectonics unit, 97–99
   qualitative and quantitative data, 70–72, **77,** 77–78, 80, **85,** 167
   quick pass through, 92
   reflecting on previously collected data, **124,** 124–125, 137–138
   Resource Activity 4.1: Data Collection Plan, 85
   for sinking and floating unit, 114–117, **118**
   for sound unit, 132
   vs. tools, 79–80
   tools for, 6–7, 37–67, 70–72
   triage of, 92
   types of, **77,** 77–78
   for uniform and nonuniform motion unit, 10
Data-driven decision making or instruction, xiii, 3, 69
DeBarger, A. H., 153
Deliberate inferences, 92–93
   analysis, 92
   going back to data, 92–93
   look behind; look ahead, 93
   Loop refresher, 92
   quick pass, 92
   summarizing inferences, 93
   triage, 92
Demonstrations, 42, 53–54, 60
Disciplinary core ideas, 4, 6, 20, 53, 139, 147, 153
   definition of, 167
   for energy conservation and transfer, 81
   for energy cycling and photosynthesis, **22, 34,** 46
   for magnetic fields and electric currents, 43
   for plate tectonics, **96**
   for structure and properties of matter, **21**
Disconfirming data, 92–93
Discussion tools, 44–46, 56
   cuing or pushing talk moves, 108
   feedback in assessment conversation, 108–112
   informal data generated by, 77
   norms for formative assessment conversations, 57–59
Duschl, R., 44

E
Earthquakes. *See* Plate tectonics unit
Electric currents and magnetic fields unit, 42–43
Elements of Feedback Loop, xiv, 6–8, **8.** *See also specific elements*
   collaborating with colleagues in use of, 130, 141–159, **147**
   connections between, 10–12, **11**
   data, 7, 69–85, 139, 167
   goals, 6, 15–35, 139, 167
   inferences, 7, 87–101, 139, 167
   in three dimensions, 12, **13** (*See also* Multiple Feedback Loops)
   tools, 6–7, 37–67, 139, 168
Energy cycling and photosynthesis unit, **22, 34,** 46
Evaluating quality of tools, 42, 66, 148
Evaluative feedback, 105–106
Everyday inferences, 90–92
Evidence-to-explanation format, 43

F
Feedback, 103–104
   definition of, 167
   strategies for, 107–113
      cuing or pushing talk moves, 108–112
      reteaching, 113

INDEX

whole-class redirect, 112–113
Feedback Loop, xiii–xiv
  closing the Loop, 103–119
  collaborating with colleagues in use of, 130, 141–159, **147**
  diagram of, **8**
  elements of, xiv, 6–8 (*See also specific elements*)
    connections between, 10–12, **11**
    data, 7, 69–85, 139, 167
    goals, 6, 15–35, 139, 167
    inferences, 7, 87–101, 139, 167
    tools, 6–7, 37–67, 139, 168
  for heat vs. temperature unit, 81–84, **82, 83**
  multiple Feedback Loops, 12, **13,** 119, 149
    leading to new goals, **130**
    for plate tectonics unit, 95–99, **100**
  overview of, 5–8
  for planning and informing instruction, 123–139
  in preservice teacher education, 157
  refining practice with, 131–135, **133, 134, 136**
  relation to *NGSS,* xiv, 153–154
  resources for, 163–164
  in three dimensions, 12, **13**
  for uniform and nonuniform motion unit, 8–10, **9–10**
  use over long periods of time, 146
Field notes, 69, 71, 72
Formal data, 3, 7, **77,** 77–78, 79, **85,** 90, 127, 167
Formative assessment, 3
  definition of, 4–5
  distinction from summative assessment, 5
  Feedback Loop and, 104–105
  probes for, 41, 148
  setting conversation norms for, 57–59
  three-step process of, 104
  tools for, 37–67
Four-corners assessment approach, 44
A Framework for K–12 Science Education, 153, 154, 167
Furtak, E. M., ix, 8–10, 15, 17, 18, 23, 30, 43, 46, 56–57, 78, 93–94, 107, 144, 148, 150, 157

G
Garmston, R. J., 145
Gitomer, D., 44
Glasser, H. M., ix, 79
Glossary of terms, 167–168

Goals in Feedback Loop, xiv, 4, 5, 6, **8,** 15–35, 139. *See also* Performance expectations
  alignment of tools with, 7, 39, 124–125, 153
    for astronomy unit, 47–52, **48, 49, 51, 52**
  for *Biston betularia* adaptations unit, 108
  cascading sets of, 130, **130**
  collaborating with colleagues in setting, 147
  connecting inferences back to, 103–104
  connections with other elements, 10–12, **11**
  correlation with student achievement, 16
  definition of, 167
  as foundation of assessment design process, 16
  grounding inferences in, 88–89
  for heat vs. temperature unit, 81
  identifying student ideas as resources for, 23–24, **25**
  identifying the gap between student learning and, 103–104, 106
  involving students in setting of, 24
  for kinetic energy unit, 37
  measurable, 17
  for natural selection unit, 126, 150
  for plate tectonics unit, 96, **97,** 98, 99
  Resource Activity 2.1: Unpacking an *NGSS* (or Other Type of) Standard, 34
  Resource Activity 2.2: Defining a Staircase Learning Progression, 35
  sequences of, 18 (*See also* Learning progressions)
  for sinking and floating unit, 114, **118**
  for sound unit, 132
  sources of, 17–18
  specific, 17
  teaching practices and, 7–8
  for uniform and nonuniform motion unit, 8
  unpacking *NGSS* into, 16, 18–22, **21, 22, 34**
Graham, S., 81, 131
Grossman, P. L., 38
Gunstone, R., 44

H
Harris, C., 153
Hattie, J., 105
Heat vs. temperature unit, 81–84, **82, 83**
Henson, K., 73
Heredia, S. C., 150

I
Identifying the gap between goals and student learning, 103–104, 106

INDEX

Inferences in Feedback Loop, 7, **8,** 87–101, 125, 139
   for *Biston betularia* adaptations unit, 112
   collaborating with colleagues in making, 149
   connecting back to original goals, 103–104
   connections with other elements, 10–12, **11**
   definition of, 7, 167
   grounding of, 88–89
      in data, 89
      in goals, 88–89
      relative to tools, 89
   for heat vs. temperature unit, 82, **83**
   importance of, 87–88
   involving students in making, 93–94
   for natural selection unit, 126, 128–129, 151–152
   for plate tectonics unit, 95–100, **96–98, 100,** 96–99
   Resource Activity 5.1: Guide to Tracking Inferences, 101
   for sinking and floating unit, 114, 117–118, **118**
   for sound unit, 132
   summarizing of, 93
   types of, 89–93
      deliberate inferences, 92–93
      everyday inferences, 90–92
   for uniform and nonuniform motion unit, 10
Informal data, 3, 56, **77,** 77–78, 79, 80, 84, **85,** 90, 127, 167
Informational feedback, 106–107
Inquiry-based science instruction, 19–20, **145,** 168

K
Keeley, P., 41, 148
Kinetic energy unit, 37

L
Lab reports, 3, 53–54, 60, 89, 124
Language of science, 4, 6, 91, **97**
Learning communities, 141, 142, 143–144, 146, 157. *See also* Collaboration with colleagues
   collaborative tool design in, 150–152, **151**
   definition of, 142, 167
   establishing meeting times for, 144
   facilitators of, 142
   formation of, 142
      engaging colleagues for, 143–144
   school-based, multiple timescales for, 152
   virtual, 131

Learning progressions, 18, 89, 92, 127
   definition of, 167
   in *NGSS,* 16, 18–22, 91–92, 148
   staircase, 18, **19,** 29, **35, 164**
      for astronomy unit in non-*NGSS* state, 26–28, **28–29**
Learning theory, 106
Lessons From Thin Air, 46
Lubkeman, K., 30

M
Magnetic fields and electric currents unit, 42–43
Michaels, S., 45
Misconceptions, 24, 71, 89, 91, 107, 168. *See also* Student thinking
   in astronomy, **28,** 49
   about heat vs. temperature, 81
   about molarity, **33**
Models, development and use of
   designing tools for, 44, 54
      for sound unit, 131–135, **133, 134, 136**
   initial and revised models, 73–76, **74–76**
Molarity assessment, **31, 33,** 112
Morrison, D., 95
Mortimer, E. F., 108
Multiple-choice questions, 9, 30–32, 40, 41, 42, 47, **49,** 72, 154, 156
   distractor-driven, 43, 154
   plus justification, 43–44, 108
   quick pass through responses to, 92
   small-group discussions and, 45
Multiple Feedback Loops, 12, **13,** 119, 149
   leading to new goals, **130**
   for plate tectonics unit, 95–99, **100**

N
Nathan, M. J., 105
National Research Council, 5, 53
National Science Digital Library (NSDL) science literacy maps, 23–24, **25, 28**
National Science Education Standards (NSES), 19–20
National Science Teachers Association (NSTA)
   NSTA Press resources, 20, 164–165
   Professional Learning Institute (PLI), 154–156
Natural selection units, 46, 93, 111, 112, 126–129, **128, 129, 151,** 154
Next Generation Science Standards (NGSS), 37, 81, 91–92, 124, 146, 153
   crosscutting concepts in, 6, 20, 167

INDEX

definition of, 6, 167
designing assessments for, 153–154
differences from *NSES*, 19–20
disciplinary core ideas in, 6, 20, 167
exploring through task analysis, 153–157, **155**
flexible connections among three dimensions of, 22
four-box structure of, 20, **21**
performance expectations in, 20–22, 167
relation of Feedback Loop to, xiv, 153–154
science practices in, 6, 20, 168
tools supporting, 42
unpacking into goals and progressions, 16, 18–22, **21, 22, 34,** 91–92, 148
Norms, 167
of collaboration, **145,** 145–146
for formative assessment conversations, 57–59
Notebooks, science, 40–41, **77,** 80, 82, 114, 124, 133

O

On-the-fly interactions with students, 45, 77, 89–90, **97, 100,** 112, 151. *See also* Everyday inferences; Informal data
Otero, V., 105
Outcome space, 40

P

Peer assessment by students, 93, 94
Penuel, W., 153
Performance expectations, 20–22, 26, 29, 39, 124, 153, 154. *See also* Goals in Feedback Loop
advantages of, 20, 22
anchoring events and, 40
definition of, 167
for energy cycling and photosynthesis unit, **22, 34**
examples of, **21, 22, 34, 65, 96**
for kinetic energy unit, 37
learning progressions and, 22
for plate tectonics unit, **96**
for structure and properties of matter unit, 20, **21**
task analysis of, 154–156, **155**
unpacking elements of, 20–22, **34,** 42, **65**
Personal-response systems, 4. *See also* Clicker questions
Photosynthesis and energy cycling unit, **22, 34,** 46

Physics by Inquiry, 9
Pitch unit, 90–92
Planning and carrying out investigations, designing tools for assessment of, 54
Planning and informing instruction with use of the Feedback Loop, 123–139
cascading sets of goals, 130, **130**
engaging with colleagues for, 130, 141–159, **147**
example of, 126–129, **128, 129**
instructional planning, 125–126.**126**
involving students in, 130
refining practice, 131–135, **133, 134, 136**
reflecting on previously collected data, **124,** 124–125, 137–138
Resource Activity 7.1: Reflecting on Previously Collected Data, 137–138
Resource Activity 7.2: Planning for Instruction With the Feedback Loop, 139
Plate tectonics unit, 95–100, **96–98, 100**
Plickers, 30, **31,** 33
Predict-explain-observe-explain tool, 44, 81, **82**
Predict-observe-explain tool, 44, **133,** 135
Preservice teacher education, 157
Probes for formative assessment, 41, 148
Professional development activities, xiii, 26, 38, 47, 49, 131, 144, 157
Project-Based Inquiry Science (PBIS) curriculum, 26
Pushing talk moves, as feedback strategy, 108, 117

Q

Qualitative and quantitative data, 70–72, **77,** 77–78, 80, **85,** 167

R

Refining practice with the Feedback Loop, 131–135, **133, 134, 136**
Reflecting on previously collected data, **124,** 124–125, 137–138
Resource Activities, 163–164
2.1: Unpacking an *NGSS* (or Other Type of) Standard, 34
2.2: Defining a Staircase Learning Progression, 35
3.1: Brainstorming Features for Multicomponent Formative Assessment Tools, 65
3.2: Evaluating the Quality of Formative Assessment Tools, 66

INDEX

3.3: Anticipation Guide, 67
4.1: Data Collection Plan, 85
5.1: Guide to Tracking Inferences, 101
6.1: Multiple Feedback Loops, 119
7.1: Reflecting on Previously Collected Data, 137–138
7.2: Planning for Instruction With the Feedback Loop, 139
8.1: Collaborative Feedback Loop Protocol, 158–159
guide to using, **164**
Resources, 20, 163–164
Reteaching, as feedback strategy, 113
Rubrics, 38, 40, 72, **77,** 94

S
Sadler, D. R., 24, 90
Science literacy maps, 23–24, **25, 28**
Science notebooks, 40–41, **77,** 80, 82, 114, 124, 133
Science practices, 4, 6, 20, 139, 147, 153
    definition of, 168
    designing tools for assessment of, 53–55, 60
        asking questions and defining problems, 53–54
        developing and using models, 54, 73–76, **74–76**
        planning and carrying out investigations, 55
        for energy cycling and photosynthesis, **22, 34**
        for plate tectonics unit, **96**
        for structure and properties of matter, **21**
Scientific inquiry, 19–20, **145,** 168
Scott, P., 108
Self-assessment by students, 93–94
Shepard, L. A., 93
Sinking and floating unit, 23, **114,** 114–118, **118**
Smagorinsky, P., 38
Small-group discussions, 45
Socrative technology, 44
Sound units, 90–92, 131–135, **133, 134, 136**
Staircase learning progressions, 18, **19,** 29, **35, 164**
    for astronomy unit in non-NGSS state, 26–28, **28–29**
Standardized tests, xiii, 3, 7, 69, 70
Stanford Education Research Laboratory, 5
Stoichiometry unit, 30
Structure and properties of matter, **21**
Student ideas as resources, 23–24, 34, **96,** 168

Student involvement in Feedback Loop, 104, 130
    data creation, 80
    goal setting, 24
    making inferences, 93–94
    tool design, 56–57
Student thinking
    anticipating student use and responses to tools, 56
    evaluative feedback on, 105–106
    formal data about, 77
    goal setting and, 23–24
    informational feedback on, 106–107
    making it visible, 40
    misconceptions, 24, 71, 89, 91, 107, 168
        in astronomy, **28,** 49
        about heat vs. temperature, 81
        about molarity, **33**
    on-the-fly interactions for learning about, 45, 77, 89–90, **97, 100,** 112, 151
    tools for assessment of, 6–7, 37–67
Suarez, E., 90
Summative assessment, 5, 9, 39, 84

T
Talk moves, 45
    cuing or pushing, 108–112
Task analysis, exploring NGSS through, 153–157, **155**
Teacher-guided whole-class discussions, 45
Teaching practices, 4, 38
    Feedback Loop for refining of, 131–135, **133, 134, 136**
    feedback strategies, 107–113
    goals and, 7–8
    learning communities and (See Learning communities)
Timperley, H., 105
Tools for Ambitious Science Teaching, 44
Tools in Feedback Loop, 6–7, **8,** 37–67, 139
    adapting of, 42–43
    alignment with goals, 7, 39, 124–125, 153
        for astronomy unit, 47–52, **48, 49, 51, 52**
    anticipating student use and responses to, 42, 56, 67
    to assess science practices, 53–55, 60
        asking questions and defining problems, 53–54
        developing and using models, 54, 73–76, **74–76**
        planning and carrying out investigations, 54

INDEX

for *Biston betularia* adaptations unit, 108–109
connections with other elements, 10–12, **11**
to create easily interpretable data, 40–41
vs. data, 79–80
data generated by, 6–7, 70–72, 80
    formal and informal data, 77
    qualitative and quantitative data, 71–72
definition of, 6, 38, 168
design of, 43–46
    collaborating with colleagues for, 147, 150–152, **151**
    discussion tools, 44–46
    involving students in, 56–57
    written tools, 43–44
evaluating quality of, 42, 66, 148
grounding inferences relative to, 89
for heat vs. temperature unit, 81, **82**
to make student thinking visible, 40
multicomponent, 34, 42, 53
    brainstorming features for, 65, **164**
    about climate change, 61–64, **62, 63**
    about magnetic fields and electric currents, 42–43
for natural selection unit, 126–129, **128**, 150, **151**
outcome space of, 40
for plate tectonics unit, 97–99, **97, 98**
Resource Activity 3.1: Brainstorming Features for Multicomponent Formative Assessment Tools, 65
Resource Activity 3.2: Evaluating the Quality of Formative Assessment Tools, 66
Resource Activity 3.3: Anticipation Guide, 67
selection of, 41–42
setting norms for formative assessment conversations, 57–59
for sinking and floating unit, 114, **114**, 118
situating in problem context, 39–40
for sound unit, 132–134, **133, 134**
for stoichiometry unit, 30, **31–33**
tool reduction, 41
types of, 38–39
for uniform and nonuniform motion unit, 8–10, **9–10**
Trauth-Nare, A., 148
Tweed, A., 104

V
Valencia, S., 38
Van Horne, K., 153
Virtual Teaching and Learning Community (vTLC), 131

W
Wellman, B. M., 145
White, R., 44
Whiteboards, 41, 46, 59, 73, **74, 75**, 148
Whole-class redirect, as feedback strategy, 112–113
Wiliam, D., 88, 104, 106
Wolfe, Z. M., x, 78
Written tools, 43–44

Z
Zekis, E., 123, 131–135, **136**